5 Acres & A Dream The Sequel

Also by Leigh Tate

5 Acres & A Dream The Book:
    The Challenges of Establishing a Self-Sufficient Homestead

Critter Tales: What my homestead critters have taught me about themselves, their world, and how to be a part of it

How To Bake Without Baking Powder: Modern and historical alternatives for light and tasty baked goods

Prepper's Livestock Handbook: Lifesaving Strategies and Sustainable Methods for Keeping Chickens, Rabbits, Goats, Cows and Other Farm Animals

The Little Series of Homestead How-Tos

Praise For

"Give a man and fish you'll feed him for a day; teach a man to fish and you'll feed him for a lifetime. In *5 Acres & A Dream The Sequel*, Leigh Tate teaches us how to fish: "How" to plan your working homestead—and not merely "what" to plan—along with not lettin' the plan plan you! This book is real-life Homesteading 101, where the value of making "a living" over making a profit is clearly put forth. And as if all that weren't good enough, Leigh gives us plenty of fish to eat in the form of top-notch homesteading "how-to's", borne from her and Dan's personal experience."
—Gary "Pa Mac" McWilliams, host of *The Farm Hand's Companion Show*

"This book opens a window into the everyday life of a homesteading couple, complete with the joys, the plans, and the dreams that you'd expect, but also the failures and the struggles that they face along the way. Rather than a textbook filled with emotionless lists of ideas or techniques, this book makes me feel like I am joining Leigh and Dan on their journey and learning what it really feels like to be a homesteader."
—Shawn Klassen-Koop, co-author of *Building a Better World in Your Backyard - Instead of Being Angry at Bad Guys*

"Those considering homesteading will get an unvarnished look at creating a self-sustaining food production system in this book. With discussions ranging from master planning the property to growing a garden and preservation, the author provides invaluable insight about the effort required to fully switch to a homegrown diet. A homestead—no matter its size—is a labor of love and that is evident in the stories the author shares about her own efforts in developing a productive piece of property."
—Kris Bordessa, author of *Attainable Sustainable: The Lost Art of Self-Reliant Living*

"Leigh and Dan are well on their way to realizing their dream of a self-reliant and sustainable homesteading life. After eleven years with a strong sense of purpose and lots of research, hard work, and trial and error, they are sharing their knowledge in *5 Acres & A Dream: The Sequel*. Leigh is a captivating writer who combines personal stories with how-to advice to create a book for homesteading dreamers, experienced small farmers, and urban dwellers who want to be responsible consumers. I'm always learning something valuable from Leigh's books and her blog to improve my own off-the-grid lifestyle."
—Margy Lutz, editor of *Off the Grid: Getting Started*

"So many homesteading texts are written by authors heavy on book-learning intelligence and light on trial-and-error wisdom. *Five Acres & A Dream: The Sequel* bucks that trend. Instead, Leigh Tate walks in the footsteps of the Nearings, whose *The Good Life* spans many decades and two states. Read along as Tate and her husband aim for self-sufficiency in both ordinary and extraordinary ways, finding the middle ground that ensures they'll stick it out for the long haul. Inspirational and educational. Highly recommended!"
—Anna Hess, author of *The Weekend Homesteader*

"This is an inspiring book to read. The dream that started "5 Acres & A Dream" is still very much alive, with lessons learned along the way that we can all benefit from. Leigh and Dan's self sufficiency is inspiring. It's great to read a book written 10 years on, and their goals of self-reliance, simplicity, sustainability, stewardship, seasonal living, and self-supporting are still the same. There's a lot of reflection and problem solving in the book, there are tons of recommendations and resources for all kinds of things, there is tons of great advice. This is not just a book, but also a homesteader's experience to learn from, and all they have encountered along the way. If you want to be inspired to follow your heart and get back to the land, read this book."
—Kate Downham, author of *Backyard Dairy Goats*

"*5 Acres & A Dream The Sequel* will take you on a journey of 2 people trying to achieve simplicity, sustainability, and stewardship on their own homestead. As you read this book you will gain insights into what it means to live as a homesteader—a life that is challenging but also rewarding and more connected to the world around you. Whether you're new to homesteading or an experienced homesteader you will learn a lot from Leigh and Dan as you follow along on their homesteading journey."
—Daron Williams, *WildHomesteading.com*

"This book made me realize that even in this go-it-alone shaping of my last farm, I can do it. I only need to reshape my path and avail myself of smarter ways to do it. I also need to accept that if what I shape is not the dream of complete sustainability, I have not failed. This book is a gold mine of information that is applicable to anyone at any stage of stewardship and sustainability."
—Terry C. Garratt, Walnetto Farm

5 Acres & A Dream The Sequel
Lessons Learned in the Quest for a Self-Sufficient Homestead

Leigh Tate

Kikobian Books
www.Kikobian.com

Copyright © 2020 by Leigh Tate
Photography copyright © 2009-2020 by Leigh Tate and Daniel Tate
Drawings and diagrams copyright © 2012-2020 by Leigh Tate

All rights reserved. No part of this publication may be reproduced, distributed, or transmitted in any form or by any means, including photocopying, recording, copying and pasting, or any other electronic or mechanical methods, without the prior written permission of the publisher, except in the case of brief quotations embodied in critical reviews and certain other noncommercial uses permitted by copyright law. For information, contact Kikobian Books, info@kikobian.com.

While this book is mostly the telling of our homesteading experiences, I have tried to include useful information for interested readers. I have made every effort to make sure that this information is accurate and up-to-date at the time of publication. If you do find errors, please be kind. They are not intentional and I am still researching and learning. In fact, some things may be out-of-date by the time you read this book. That being said, please do, thoroughly research information that is new to you before utilizing it. Please make certain it is correct, appropriate, and applicable to your situation.

This book was built, from the ground up, with open source software on an open source operating system: Xbuntu Linux 18.04 (Bionic Beaver), LibreOffice Writer 6.0.7.3, Gedit 3.28.1 text editor, Gimp 2.8 photo editor, Zim 0.68-rc1 desktop wiki, and Scribus desktop publisher versions 1.4.8 and 1.5.6. Also used were open source fonts Lobster Two, EB Garamond, EB Garamond SC (SIL Open Font License Version 1.1.), and Liberation Sans (GNU General Public License v.2).

ISBN 978-0-9897111-4-2

Kikobian Books
www.kikobian.com

To my Dad,
who taught me to love and respect nature.

Contents

Acknowledgments		xi
Introduction		1
Chapter 1	The Dream: Is It Still Alive?	3
Chapter 2	Reassessing Our Goals	9
Chapter 3	Reevaluating Our Priorities	21
Chapter 4	Fine-tuning the Master Plan	31
Chapter 5	The Transition Phase	53
Chapter 6	Food Self-Sufficiency: Feeding Ourselves	57
Chapter 7	Food Self-Sufficiency: Feeding Our Animals	85
Chapter 8	Energy Self-Sufficiency	101
Chapter 9	Water Self-Sufficiency	131
Chapter 10	Resource Self-Sufficiency	143
Chapter 11	Discouraging Things	163
Chapter 12	Distractions	175
Chapter 13	Toward Keeping a Balance	179
Conclusion	A Sense of Purpose	189
Homestead Recipes		
Fiesta Cornbread		193
Oven-Fried Okra		194
Probiotic Ice Cream		195
Nutrient Dense Bone Broth		196
Apple Pectin		197
Spicy Fig Jam		199
Heavenly Chèvre Cheesecake		200
Appendices		
A. Resources		230
B. Pasture Rotation: 3 Models		207
C. Polyculture Forage, Hay, and Cover Crop Lists		209

BIBLIOGRAPHY	215
INDEX	217
INTERESTED IN MORE?	245

Acknowledgements

It is sometimes thought that authors write solely for the pleasure of writing; that their need to create with words transcends all other reasons. While there is perhaps some vague truth in this, I think most of us are prompted by other motives. For me, it has been a request from readers for more. That is the reason you now hold *5 Acres & A Dream The Sequel* in your hands.

Many people helped shape this book in ways large and small. To each of them, I want to extend a heartfelt thanks. Thank you to everyone who emailed me or commented on my blog. Your feedback on both my books and my blog helped me understand what your needs are and where your interests lie. Hopefully, this work will be another encouragement in your life journey.

Thank you to my blog readers whose feedback helped me fine-tune my latest Master Plan: Deborah, Stephanie, MaryP, Retired Knitter, TaniW, Woolly Bits, Ed, Unknown, Boud, Lynn, wyomingheart, Florida Farm Girl, Susan, Debbie-MountainMama, Lady Locust, Ann, Toirdhealbheach Beucail, Kris, Ron Clobes, Renee Nefe, Mike Yukon, Toni (in Nigeria), City Creek Country Road, Helsyd, Sandi, Cockeyed Jo, Fiona, Chris, and Val Champ.

A special thanks goes to Robbie Auman, Dino Garnett, and Elaine Shanks for taking the time to read my almost-final draft and offering suggestions for improvement. Your enthusiasm and honesty have been a blessing.

To my anonymous editor, a huge thank you. I claim any errors or mistakes as solely my own.

Lastly, a very special thanks has to go to my husband Dan. We are two very different personalities sharing the same life, so his perspective confirms, contrasts, clarifies, and balances my own. My writing is all the richer for it.

Introduction

When I first published *5 Acres & A Dream The Book*, I never dreamed it would do as well as it has. Nor that I would hear from so many people telling me how encouraging that book has been to them. When I look back through it, however, I am aware of how much our homestead has changed. Between that and the encouragement I receive to continue writing, I felt that it was time to share the next part of my husband Dan's and my story.

In my first book, I describe Dan and myself as an empty nest couple. Now I would describe us as being in early "retirement." My readers should take that term with a grain of salt because our idea of retirement has little to do with what most of the modern world assumes. While it means that Dan no longer "goes to work," it doesn't mean we are now living the life of leisure that most folks associate with retirement. Rather, we are working more, but loving it more.

As "retired" folk, our retirement income is modest, less than the working income to which we were accustomed. It came about sooner than we expected, a tale that I will tell in the following chapters. What I will say here, is that because we have been diligent toward our goal of self-reliance, the transition wasn't as difficult as it might have been. We were geared toward greater self-sufficiency from early on and experienced several periods with no income. Each of these times was difficult but served to prepare us for the fixed income lifestyle that many retired people have to struggle with. As it is, I have no complaints about our current way of life.

The format for *5 Acres & A Dream The Sequel* is similar to *5 Acres & A Dream The Book*. Chapters and topics parallel one another, and each chapter begins with a quote from one of my books. Hopefully, these quotes will set the stage for you, dear readers, so that no lengthy summary is needed to jog your memory as to the challenges we were facing then. Or if you haven't read my first book, you can still get the gist of our whole story because it hasn't been the challenges that have changed, it has been what we've learned and how we've met them.

Even so, you may be aware of some gaps in that story, because after I wrote *5 Acres & A Dream The Book*, I wrote *Critter Tales*. Where *5 Acres & A Dream* covers all aspects of our homesteading journey, *Critter Tales* focuses on our experiences and learning curve with homestead livestock keeping. Although never meant to be an extension of my first book, it helps fill that gap between my first work and this one.

I also want to mention that my citations for the quotes from my earlier books will be somewhat atypical of most style guide recommendations. I'm taking this liberty because I've already told you I am quoting myself. If I quote anyone else, I'll give you a heads-up on who said it and where.

One more comment before you get on to chapter one. It has been suggested to me that the *5 Acres & A Dream* book series would be greatly improved if the photographs were in color. I agree. What keeps me from printing in color is cost. Printing cost alone would add an additional $15 to the price of each book, with standard publication-based percentage fees added on top of that. I want my books to be as affordable as I can make them, which means printing in black and white.

Addendum

Earlier this year, the COVID-19 pandemic descended upon the world, and like many other nations, mine went into lockdown status. The text for *5 Acres & A Dream The Sequel* was already written by that time, and the book was deep into preparation for print. Consequently, you will find no mention of the pandemic within the years covered in this sequel. It's a significant event, however, so this introduction seemed the best place to comment on it.

How did the pandemic affect us? It didn't. Because of our lifestyle, our lives continue the same as before. But I know that hasn't been the case for everyone, and that there are many people whose sense of reality has been shaken. If you're one of them, my heart goes out to you.

If you're looking to take back some sense of control over your life, then I hope this book is an encouragement to you. I hope it can help you develop a new normal of your own making, with a greater sense of purpose and inner freedom. You don't need large acreage or a lot of money to start a garden. Or to simplify your diet. Or your lifestyle. You just need to take a first step. We all start the journey at the same place—the beginning.

Chapter 1

The Dream: Is It Still Alive?

"There is no specific point that either my husband Dan or I can pinpoint as being the birth and definition of our dream. . . . Rather, it has been an attraction to a way of life, to what we thought would be more fulfilling and personally more productive than the typical lifestyle of our culture."

"The Dream," 5 Acres & A Dream The Book *(p. 3)*

If you were to ask me what has changed on our homestead since I wrote *5 Acres & A Dream The Book*, I would tell you, "a lot." At a glance, everything looks quite different. The house is now blue and features a bay window overlooking a large open front porch. At the end of the driveway stands the goat barn, exactly where our old coal barn used to be but newer and fresher, with my hand-painted barn quilt gracing the hayloft doors. Behind it, the original chicken coop and goat shed has been expanded into a workshop for Dan and sports a new metal roof. Next to that is the chicken coop Dan built several years ago and the enlarged chicken yard. The two old oaks that I loved finally died and became firewood. In their place, three solar panels now stand at attention.

Above: Our original driveway (2009).
Below: Our driveway today (2020).

We called the original barn (above) the "coal barn" because it once housed coal to heat the house. The new barn (below) has the same footprint, but with a hayloft.

    The biggest changes, however, are not what you see when you look around. The biggest changes are in how we are learning to view ourselves and our relationship with our homestead. There is a long litany of "failed" experiments to go along with that change, each resulting in a flurry of new research, not to mention soul-searching. Yes, there have been times we've questioned what we're doing here and whether it's worth it. More than once we have discussed walking away, but that discussion doesn't last long. Besides the obvious question of, "what else would we do?" there is an inner conviction that this is how we are supposed to live.

Above: The house when we first arrived in 2009.
Below: The house after extensive repair and upgrading.

*Above: The outbuilding we first used as chicken coop and goat shed.
Below: The same building expanded to become Dan's workshop.*

The more we interact with the natural world around us—physically, mentally, emotionally, and spiritually—the more we understand ourselves to be a part of it. Our life's work is to conduct ourselves in such a way that our five acres of earth can be its best self. And that begs the question of how. How do we function as part of our homestead ecosystem?

That idea is counter-cultural to modern thinking. Modern thinking tends to view humans as an environmental problem. It's true, humans are extremely destructive creatures, but if humankind is truly "The Problem," then someone or something got it wrong. Either God was wrong in creating us in the first place, or evolution was wrong by selecting us to become the dominant species. What is true, is that there is an extreme disconnect between modern culture and the natural world. Urbanization and technology are leading people away from nature. That influences how they see it, how they think about it, and what they want to do with it.

Unfortunately, today's high esteem for technological advancement and gadgetry is a blind spot in the modern point of view. Problems are recognized, but causes are ignored. Research is based on reductionist science rather than the whole, and the recommended solution is always to throw more technology at the problem. But haven't they noticed things are only getting worse?

> *"We longed for a simpler life, a life that gave us a sense of purpose, appreciation, and satisfaction with what we do and how we do it. We wanted a lifestyle that relied less on consumerism and more on our relationship with the natural creation and its gifts."*
> "The Dream," 5 Acres & A Dream The Book *(pp. 3-4)*

The longer we homestead the more this is true, and the world's way continues to become less attractive. Consumerism is certainly less appealing, in part because we prefer what we can grow and produce ourselves. But also, because of the increasingly poor quality of commercial goods being produced nowadays. Food from the grocery store has no substance; no real flavor. Construction materials are becoming smaller, flimsier, and contain more flaws. The metal used in tools and equipment is either lighter gauge or replaced with plastic, which eventually cracks and breaks. When faced with the choice between buying such offerings or doing something else, we increasingly have chosen to do something else.

Sometimes that something else is doing without. We've learned that we can truly live with less. It's not that we're settled for less, it's that we realize we can be content with less. The old saying "use it up, wear it out, make it do, or do without" is good advice. The interesting thing is that when we don't continually rely on buying solutions to problems, the brain somehow switches into its creative mode. By casting about for alternatives, new ideas present themselves. I'd say we've accomplished more with fewer financial resources than we thought possible. And we're happier for it! Because of that, I can say with certainty we are indeed living our dream.

How have these changes impacted our goals? More on that next.

Chapter 2

Reassessing Our Goals

"Understanding what we wanted from our land and defining how we would go about it was an important first step. In fact, it was crucial in helping us define our primary goal of working toward self-reliance. This, in turn, continues to give us direction and helps us set our priorities and evaluate choices as they present themselves."
"Defining Our Goals," 5 Acres & A Dream The Book *(p. 25)*

My readers may well ask if this statement is still true. Is self-reliance still our primary goal? Does it still give us direction? Does it continue to serve as a means to establish priorities and help us in our decision-making?

To answer those questions it's important to understand that by self-reliance, i.e. self-sufficiency, we do not mean cutting ourselves off from the world. This is a common misconception about homesteaders and one that is often used to criticize them. As Dan and I define self-sufficiency, it means "not being eternally dependent upon the consumer economic system

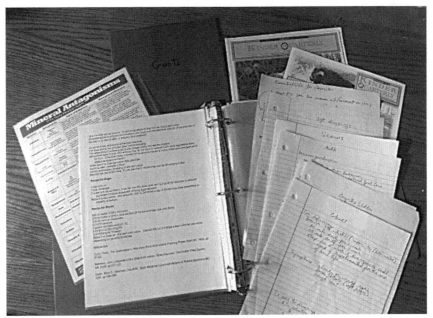

I think of self-reliance on many levels, including information and resources. Having hard copy notes means I'm not dependent on a digital device to store information.

to meet all of our needs. It means learning to rely more on ourselves, on our willingness to work, learn, and adapt, on our ability to problem solve, and on our lifestyle choices. It means relying less on external resources, especially those that must be purchased in order to live."[1]

As our nation's ideological divide continues to widen, our goal seems as relevant as ever. I'm writing this with the 2020 presidential election just a year away, so it's almost impossible to think that the current clash won't escalate as the power struggle becomes more forceful. With common folk caught in the political crossfire, I believe that our lifestyle choices will make the difference between treading water or drowning in the potential chaos and its consequences. But there's more to it than that. Our homesteading lifestyle provides a profound sense of purpose. Not in terms of preparing to survive the upcoming social upheaval or a potential economic collapse, but in terms of the life-meaning it gives to us as humans. There is something profoundly satisfying in working toward a less complicated, less wasteful, and environmentally responsible lifestyle.

In *5 Acres & A Dream The Book*, I discussed three principles that represent the core of what we're working toward: simplicity, sustainability, and stewardship. These continue to define how we're working toward our goal of self-reliance. That hasn't changed. If anything, they become more relevant now that Dan is no longer in the workforce. These principles are key to keeping things manageable as we get older.

Simplicity

> "A simplified life is a life not complicated by the pressure of being, doing, or having whatever the latest social trends dictate. We do not want to be slaves to our jobs, nor to accumulate as much money, material wealth, and financial security as we possibly can, nor have as much leisure and "fun" as we can get. In a nutshell . . . we refuse to compete in the rat race of life."
> "Defining Our Goals," 5 Acres & A Dream The Book (p. 21)

Simple soups, stews, and salads are easier to prepare and quicker to clean-up.

In our early years here, simplicity was elusive. This wasn't because we were still entangled in the rat race of life, it was because we were working to establish our homestead while Dan worked full-time. We didn't have social pressure, but we created our own by pushing ourselves to get it all done. Yesterday.

We worked on two fronts simultaneously. One was infrastructure. We needed livestock shelters and fencing. We needed a place to store tools, equipment, feed, and firewood. We needed to repair and refit the house for energy efficiency, wood stoves, food storage, rainwater collection, etc. To achieve all of that we needed materials, supplies, equipment, and tools, plus gasoline and electricity to power the equipment and tools. Even though we wanted to rely less on the consumer system, those needs kept us dependent on it. We understood and accepted this as part of working toward our goal, but sometimes it was frustrating.

The other front was establishing a homegrown food supply for both ourselves and our critters. We worked to establish vegetable and herb gardens, fruit trees and bushes, field crops, pasture, and areas to grow hay. I did extensive research, and we learned a lot through trial and error. If we achieved any simplicity in those first years, it was because we simplified our diet. We chose to eat more of what we could produce ourselves and less of what we could buy at the grocery store.

Ten years later we are finally getting our infrastructure in place. However, simplicity is not coming automatically. Just *wanting* to simplify our lives doesn't help us cross the bridge to reality. We must continually remind ourselves that it isn't a race to complete all the tasks on a checklist, rather, we are working toward a lifestyle. A lifestyle that will keep us engaged in the process of working toward self-reliance.

Working smarter, not harder isn't instinctive. It's something that has to be figured out as one goes along. Growing a garden, for example. Our grandparents controlled weeds with backbreaking hoes because that's the way it was done. The invention of tillers must certainly have seemed like tremendous progress, but today's gardeners know that mulch makes weed control even easier. It's a simpler solution with the benefit of feeding the soil rather than the tiller.

Something else we've learned is that simplifying a thing on one level doesn't mean it can be transferred to another. A thick layer of mulch may make it easier to control weeds in garden beds, but what if I want to grow an acre of wheat, pumpkins, or corn? While it's certainly possible to use the same method on an acre or more, finding that much mulch isn't so simple. The reality is that larger plots of food require different methods. The only answers we could see then were from modern conventional farming: plowing, tilling, and herbicides. But these are a far cry from the more natural ways we wanted. While we were physically closer to nature, the simplicity we were looking for eluded us. Isn't there a better way, we would ask. Yes, there is, and I'll share what we've learned in my chapters on food and resource self-sufficiency.

Sustainability

"Sustainability requires that we not use up what we have to the point where there is no more."
   "Defining Our Goals," 5 Acres & A Dream The Book *(p. 21)*

Livestock lend themselves to a sustainable supply of eggs, milk, and meat.

Considering how trendy the word "sustainable" has become, my above definition might seem over-simplified. Nowadays, its meaning can differ greatly depending on how it's used. We hear of sustainable energy, sustainable living, sustainable economics, sustainable development, and sustainable design, for example. Is sustainable energy the same as sustainable economics? Is sustainable agriculture the same as sustainable development? Or does "sustainable" have different meanings in different contexts?

I researched enough definitions of these terms to conclude the latter. And that points to not making assumptions about the meanings of words outside of their contexts. This is why I need to make sure that my definition is consistent with my goal. As I contemplated my definition of sustainability, I found two more at *Dictionary.com* that I liked.[2]

Sustainable
1. Capable of being supported or upheld, as by having its weight borne from below.
2. Pertaining to a system that maintains its viability by using techniques that allow for continual reuse, such as sustainable agriculture.

Do those definitions describe our homestead? Have we been taking action toward meeting that goal? In the beginning, we focused on food, water, and energy, and yes, we've taken steps in those areas. Now I would add other resources, which I'll discuss in my chapter on resource self-sufficiency.

What I have come to understand in recent years, however, is that the foundation of our sustainability is our soil. Every decision we make and technique we use needs to keep it going and growing. The problem has been that our soil is regionally poor, exhausted from centuries of farming and abuse. Building soil in a garden bed is one thing, over acreage it's another. Understanding how to do that was slow in coming, and I'll discuss what we've learned in detail in upcoming chapters.

Stewardship

> *"Stewardship evokes a sense of responsibility . . . It implies the supervision or management of something entrusted to one's care. It implies not only responsibility but also accountability. We believe that one day, we will be accountable for how we lived our lives and for what we did with the things in our possession."*
> *"Defining Our Goals,"* 5 Acres & A Dream The Book *(pp. 23-24)*

The longer we homestead the more this rings true. But because we don't truly understand natural processes from nature's point of view, we make a lot of mistakes. Even though our plans and motives have been well-meaning, we've had a lot to learn. Two of the most important lessons have been:
1. We aren't as smart as nature.
2. We don't control anything.

Ironically, we aren't supposed to be in control. Our place in the scheme of nature is to observe and respond when appropriate to meet the needs of our land, our animals, our vegetation, and our soil. In the past, I've written that we have considered ourselves to be partners with our land and our

Protecting livestock from predators is part of our stewardship. Opossums, for example, will kill and eat poultry and eggs.

livestock, but the truth is that we are their servants. If we can understand what's required, then we can meet the need. It's all too easy to formulate our plans and then try to implement them according to a personal measure of success. Unfortunately, our idea of success is not nature's success. Humans strive for productivity, nature strives for balance. Our proper place and function in Creation are to work toward that balance. The problem is that we humans always tend to tip the balance toward ourselves.

If we truly want to improve the health of our plants, animals, soil, and ultimately ourselves, then we must do things nature's way. This is how the system was set up in the first place. Centuries of science and technology have never been able to improve on that. If anything, they make things worse. Not in terms of leisure time and the ability to possess stuff, but in terms of the environment and our planet's ecological health. This is why stewardship remains one of Dan's and my essential goals.

In addition to simplicity, sustainability, and stewardship, we've worked toward other goals as well.

Seasonal Living

Living seasonally is inseparable from a lifestyle that depends on planting and harvesting. Even so, as a goal it's been harder to attain than we first thought.

"As the big, one time projects are completed, we look forward to a time when our lives will no longer be dictated by the next self-sufficiency project on the list. We look forward to developing a seasonal routine, one based on the rhythms of spring, summer, fall, and winter. That seasonal lifestyle, which is a hopefully simpler lifestyle, is one of the reasons we were drawn to the homesteading life in the first place."

"Where Do We Go From Here?"
5 Acres & A Dream The Book *(p. 185)*

When I wrote that, we were still busy trying to repair, upgrade, and build. At that time the seasons were little more than calendar holidays and what the thermometer said. We didn't appreciate them, because they required that we take an impatient pause in the middle of a project to plant lettuce and peas before the summer heat arrived, pick tomatoes before the first frost, or chop firewood before winter's cold descended upon us. At that time, the seasons were dictators that required certain activities be done at particular times of the year.

I think it was eating our homegrown food that imprinted us with a sense of true seasonal living. Many of you can relate to that. By choosing to eat foods grown in the local season, one begins to hold hands with a rhythmic pattern of eating and growing. I don't have to tell you that nothing tastes as good as spring's first salad. Or summer's first bite of fresh blueberry pie. Or the season's first oven-fried okra. Add to those other seasonal foods such as eggs and milk, and our diet became the heart of a seasonal way of living.

As we made progress on our infrastructure, we found ourselves focusing more on what the season required, rather than on some pressing project to be done. This, in turn, makes project planning easier, because now we plan activities according to the seasons.

Becoming more seasonal in our lifestyle has changed our concept of time too. It used to seem odd to me that the beginning of a new year is in the middle of winter. It was very interesting, then, when I read Eric Sloane's *The Seasons of America Past* and learned that the old agrarian calendar began the year on March 25.[3]

The roots of this go back to the ancient Hebrew, Canaanite, and Babylonian calendars, where the year started in the modern late March. The early Romans celebrated the new year on March 1st, but this was changed in 45 BC with the Julian calendar. It set January 1st as the beginning of the new year. The best I can figure out is that this was done because the month's namesake, Janus, was the god of new beginnings. When Pope Gregory XIII had the calendar revised in the mid-1500s January remained the first month.

The Gregorian calendar is the one we still use today, but March 25 remained the start of the new year in parts of Europe for a long time. England was the last holdout until 1752. According to Eric Sloane, early European American farmers continued to use March 25 as the beginning of their new year, as documented by many old farm calendars, almanacs, agricultural manuals, and personal diaries.

With March as the first month, the quarters of the year make sense because the seasons correspond with the work for that time of year.

March, April, May – planting
June, July, August – growing
September, October, November – harvest
December, January, February – hearth

As you can see, the rhythm of the agrarian year is set by a relationship with the seasons and with the land. As our lifestyle became more and more tied to the land, this view of cycles and beginnings became more natural. In recent years it has become one of the ways we set priorities, something I'll discuss at length in the next chapter.

There is one more goal we contemplated early on. Working toward this one has been a bit more complicated, however, and our view of it has not only changed with experience, but with our growing older.

Self-supporting

"Although there were obstacles, we entertained the idea of becoming career farmers. Besides not having the start-up capital, our biggest problem was that we had no experience in farming. . . . (also) neither Dan nor I have a business mindset. . . . As much as we wanted to make our living from our land, . . . we had to ask ourselves how this fit in with simplifying our lives."
"Defining Our Goals," 5 Acres & A Dream The Book (p. 19)

Dan has struggled with the desire to become self-supporting more than I have. His job as an over-the-road truck driver kept him away from home for long periods at a time, which meant that progress on the homestead pretty much came to a standstill when he was gone. I did what I could by maintaining the more routine aspects of our lifestyle: daily chores, the garden, critter care, food preservation, etc. Key projects such as fences, outbuildings, house upgrades, and repairs, took 100 percent of Dan's time while he was home. Progress was understandably slow, which was frustrating. On top of that, he didn't enjoy his job. He wanted to be home full-time to get more done. But there were obstacles.

I mentioned several obstacles in the above quote, but I think the biggest was uncertainty. The list of things we didn't know far outweighed our confidence that we could successfully tackle that list. Besides lack of experience and start-up capital, we would have to learn to navigate whatever government requirements, regulations, and red tape would be necessary to make a living from our homestead. While some people might relish the challenge, for us it was overwhelming, especially because of Dan's job.

If Dan had worked a standard day job, he could have started a farming venture part-time, learned the business, and worked it into a full-time job.

A full-time trucker, however, typically has about 36 hours of hometime for each 70-hour workweek. This is enough for a legal reset of their logged driving time, but barely enough time to rest, do laundry, and get ready for the next trip out. It offered no extra time for projects.

The ability to pay the mortgage, utility bills, and other ordinary living expenses came from Dan's driving income. I know some folks think that truck drivers make a lot of money, but driver wages have plummeted over the years. At one time, a driver who owned his own rig could make over $100,000 per year if he was willing to be away from home 95 percent of the time. A company driver might earn up to $55,000, if he or she was willing to be on the road most of the year. But those salaries translated to very little hometime. Since about 2008, driving salaries across the board have bottomed out and stayed there. Simply quitting driving to jump into farming with zero income wasn't an option.

During this time I managed to find a small income stream through writing. While Dan's salary paid for ordinary needs and expenses, everything I earned was invested in tangible assets for building the homestead. Thanks to you, my readers, book royalties have enabled us to buy equipment, tools, building materials, and our critters. It's even enabled Dan to take a few leaves of absence to make progress on the homestead.

Another area in which we've had a success of sorts is our Kinder goats. We make enough money through goat sales to pay for whatever livestock feed we don't grow, plus supplies. Some might not think much of just breaking even, but to me, it's a success because our livestock are paying for themselves. They are self-supporting.

When Dan had his accident (more about that in chapter 12, "Discouraging Things"), everything changed. I could still manage the seasonal chores of day-to-day homesteading, but it put all projects on hold for a time. Even so, we still had monthly bills to pay and necessities to buy. I didn't make enough through book royalties to cover all that, and because he couldn't work we explored other options.

The immediate options were either disability or early social security retirement. After a visit to the area social security office he chose early retirement because it was the quickest and simplest solution.

Currently, we pay monthly bills and buy necessities from our social security deposits. But is social security really secure? That's another kettle of fish, but I'd guess it's just as secure as any of the other retirement schemes out there. Still, our self-reliance goal requires us to ask if there is a way we can meet our needs without it. No matter how self-sufficient we become, there are some things for which we will always need money.

Now that we are officially retired, Dan and I still discuss becoming self-supporting. But is it achievable? I have to confess that as we get older we're slowing down. There is less enthusiasm and energy to build a profitable business from scratch. In truth, we're more interested in making a living rather than making a profit.

It's important to point out that there is a difference between the two. That's hard for many people to understand because our economic system is geared toward economic growth. Our culture is geared toward attaining and securing monetary wealth. Dan's and my underlying motive is different. We aren't looking for material affluence and leisure; we're looking for a sense of meaning and purpose. Call us odd fish, but the quest to accumulate money doesn't give us that. We find it in a hands-on lifestyle that requires us to partner with nature in meeting our fundamental needs. And that, my friends, requires one of the most elusive of human qualities—contentment. If we can maintain contentment and not be seduced by the lure of making oodles of money, then yes, I think becoming self-supporting is an achievable goal. With Dan at home full-time and most of the one-time building and repair projects done, we can focus on new avenues for a modest income.

These six goals: self-reliance, simplicity, sustainability, stewardship, seasonal living, and self-supporting, make up the lifestyle toward which we are working. That hasn't changed, but can we measure our progress? To make the primary goal of self-reliance more manageable, we initially organized it into three categories: food, water, and energy self-sufficiency. Several years later we added a fourth—resources. I'll detail our progress on these in the upcoming chapters.

Notes

[1] Leigh Tate, *5 Acres & A Dream The Book* (Kikobian Books, 2013), 22.
[2] "Sustainable," *Dictionary.com*, https://www.dictionary.com/browse/sustainable, accessed October 25, 2018.
[3] Eric Sloane, *The Seasons of America Past* (Mineola, New York: Dover Publications, 2005), 32.

CHAPTER 3

REEVALUATING OUR PRIORITIES

"Once the excitement settled, the amount of work that needed to be done was overwhelming. It was all too easy to become distracted with details, such as the style of chicken coop we wanted or what breed of goats we should get. The thing that helped was continually keeping our primary goal in mind, that of becoming as self-sufficient as we are able. What was important was determining the steps we would need to get there."

"Setting Priorities," 5 Acres & A Dream The Book *(p. 31)*

I've never claimed to be a good multitasker. If anything, I'm the opposite. I tend to be overly focused on whatever task or project is currently on my radar. Dan is pretty much the same way. The problem is that when there's much to do it's easy to become overwhelmed. Even though we have completed most of our big one-time projects such as fencing, building the goat barn, remodeling the kitchen, and replacing the old windows in the house, there remains a never-ending stream of needs, problems, and ideas presenting themselves.

Some of these are maintenance projects such as routine fence repair. Others are unexpected repair projects, like discovering a leak in the pantry roof. We've learned that some needs are more obvious than others. A broken window is immediately obvious; a mineral deficiency in the soil isn't. Or perhaps an idea that seemed good at the time hasn't been as practical as we hoped. Sometimes these can be tweaked, such as our rainwater collection filters. Some projects are set aside, such as building a pond. Some new projects come from experience or our ongoing research and learning. It may be a matter of stumbling across someone else's brilliant idea or the result of trial and error.

Deciding what to do next isn't as straightforward as it ought to be. It's easy to get caught up in the bumper car approach, i.e., tackling whatever one bumps into. But bouncing randomly from one project to another tends to be neither organized nor systematic toward making progress. That's why keeping a primary goal in mind is important. Even so, when the project list is long there is often no clear-cut path for making progress.

As we've completed many of our one-time establishment projects, our thinking has changed. While we still have goals and plans, somewhere along the way our focus shifted from the next project on the list to what the season demanded. We shifted from pushing ahead with a linear mindset to the cycle of the seasons. Even so, we still have a lot to do.

Cleaning the wood stove chimney.

So how do we prioritize our project list? The first step is to set aside time to plan. We write a to-do list, and then we organize it by categories: seasonal, self-reliance, maintenance, ongoing projects, repair, and anything demanding immediate attention.

Seasonal projects will typically revolve around food production. But they also include tasks such as cutting trees for firewood and cleaning out the gutters, chimneys, and wood stoves.

Log splitting for firewood is another seasonal chore.

Self-reliance projects include anything that will help us toward that goal, things like expanding our rainwater collection, digging a root cellar, building a greenhouse, or installing a small solar energy system.

Maintenance. These jobs often revolve around machinery, such as tune-ups and oil changes, but also include other things: mucking out the barn and adjusting gates. For some reason, our gates tend to sag so that the bolts don't slip easily into their latch. Their routine adjustment is a maintenance chore.

When we have problems, we often ask, "is there an alternative?" The simple homemade gate latch shown on the right solved the problem of replacing and adjusting purchased latches.

Ongoing projects include chipping wood, mowing the lawn, etc. These aren't specifically seasonal, but routinely need to be addressed.

Repair can be vehicle and machinery repairs, or things like addressing that leak in the pantry roof. Projects in this category can be urgent, important, or those that can wait. A tree falling on a fence will urgently need tending to if we don't want our goats invading the neighbor's garden. Replacing aging fence braces can be worked in as time allows.

Trees falling on fences take priority over everything on the to-do list.

Daily chores are not included on the project list because they don't have to be planned for and prioritized. They are the foundation of our routine. They open and close the day, like bookends for everything else we need to do.

After our list is written, the next step is to prioritize. We do this with a checklist we've worked out over the years.

Food first. Food is our top priority, so anything food-related, whether for us or our critters, must come first. This is particularly true of planting, harvesting, and processing, because they are dependent on our growing season. None of these things can wait.

Exceptions do exist, for example, digging a root cellar. For many years I thought I could do without one because I can store canned or dehydrated foods in my pantry. Milk, cheese, and eggs are stored in the refrigerator. Winter root crops can be stored directly in the garden under heavy mulch. For some things, however, especially winter squashes and sweet potatoes, my pantry gets too warm when the weather is hot. While a root cellar would help with food storage, it's a project that will require location planning, materials acquisition, and a means to dig it out. In the meantime I make do, until building a root cellar is feasible.

How pressing is it? Are there temporary alternatives? This requires thinking through the consequences of not doing something. A leaky roof needs repair sooner rather than later to avoid extensive damage. This was the case when Dan discovered the leak in our pantry roof. Yet we also wanted to finish the goat barn before winter weather set in. A temporary solution for the leaky roof was to tarp it until the barn was finished. Then he reroofed the pantry.

The tarp stopped the leak until the pantry could be reroofed.

Can we get materials? Can we afford them? Is there an alternative? Getting what we need has become harder over the years. Most of our local businesses aren't geared toward the kind of lifestyle we live, and if they are, their prices are geared toward the hobbyist. Chain stores now seem to stock less on the shelves because they rely more on internet sales: order it online and have it shipped free to the nearest local store. After that, it's finding it on Amazon, eBay, or an independent seller website. Shipping may or may not be free, and high shipping is always a deal-breaker. All of these things have prompted us to look for alternatives.

One alternative to buying materials is making them ourselves. When we first calculated the cost of building the goat barn, the sum was discouraging. Dan commented that if he owned a sawmill, he could take advantage of the numerous pine trees on our property. They are large and old, and many of them were beginning to fall on their own. We hated to see them go but also didn't want them to go to waste. Couldn't we use them as lumber? We turned to Craigslist to see what we could find and there it was; the prefect small portable sawmill at the perfect price. That little sawmill has more than paid for itself.

Harbor Freight Central Machinery portable sawmill.

Time versus money. I suspect almost every one of you is familiar with this dilemma: Do I save money and do it myself, or do I save time and buy it? That may mean buying the item outright or buying the tools and supplies to do it myself. Or it may mean hiring someone to do the work. What's the best choice? I don't know about you, but sometimes one wins out, sometimes the other. There is no right or wrong answer.

When it comes to prioritizing, we need to know if we have the materials, supplies, and tools for a project, or the means to obtain them. For tools, we've both purchased and rented. If an expensive tool will only be used for one project, renting makes more sense. If the tool will be useful for other projects, purchasing is usually preferred, as long as we have the money for it.

We don't buy on credit which helps with project prioritization too. If we don't have the cash, we must ask, "can we accomplish this without buying anything?" Sometimes we can, and other times we shelve the idea for the time being.

Potential maintenance. When weighing the merit of a project, it's important to consider its long-term needs. What resources will be required to keep it going? Once in place, what long-term physical ability and energy levels will it require? Will it remain manageable as age slows us down? Are there alternatives? Are there simpler ways to accomplish the same objective?

There are many good ideas for homesteaders out there, but they often require significant time and energy to maintain. Brewing compost tea or making biochar are two things I considered. They are excellent ideas, but would I be able to balance the time they required with everything else? Would I be able to provide all the components myself or must I continually buy something to maintain them? The answers to projects like this are subjective because there are many individualized factors to consider. We chose other soil building methods because they worked better for our homestead goals and routine.

Short-term versus long-term projects. Another way to think of this is one-time projects versus ongoing projects. Building the new goat barn felt like a long-term project because it spread itself out over several years. But in the grand scheme of things, it had a decided beginning and end. Now that it's finished, we don't think about the time it took to build it. Pasture improvement, on the other hand, takes years to achieve. It's truly a long-term project.

Ideally, long-term projects should work themselves into the seasonal routine. In choosing to embark on such projects, the question becomes, what season is the project best suited to? What time of year will we have the most time? It doesn't always work out according to the ideal. For example, more than once we've found ourselves replacing windows in the winter!

Building the goat barn has been our most ambitious project so far and took almost two years to complete.

Sometimes the ideas, energy, and resources for a project simply present themselves. Sometimes we recognize a need, but what to do about it remains elusive. Each time we discuss it, we can't decide. Then suddenly, it becomes clear. Everything falls into place and we can proceed.

This has been especially true of our big one-time projects. The goat barn is an example. It was prominent on our first master plan and has remained so on every revision since. But we couldn't seem to figure it out. I can't tell you how many plans we drew up: six, seven, eight? Yet we couldn't get beyond that. We had lists of questions that needed to be planned out, but we kept drawing blanks. Or other projects overshadowed it. Honestly, for a long time, it seemed as though a proper barn was simply pie in the sky. After we got the sawmill, everything else fell into place.

Once our projects are prioritized, working our way through them is pretty informal: "what should we focus on this week?" and "what are your plans for the day?" The season and weather are two important factors when it comes to choosing the day's or week's projects. The growing and harvesting seasons have more demands, which means less time for everything else on the project list. Weather, of course, determines the day.

Analysis and revision. Periodically, we review, evaluate, and update our project list. We don't expect to accomplish everything on it, but we do revise it routinely. Completed projects are crossed off and new ones added according to the season or because something new has presented itself. Then we take the list back through the process I described above.

Ten years into homesteading, I can tell you that the list never gets shorter or goes away. But I can tell you that by taking the time to analyze, evaluate, and plan, we can keep our goals manageable and our priorities straight. It's a tool that works well for us and sets the pace and tone of our seasonal rhythm and work routine.

CHAPTER 4

Fine-Tuning the Master Plan

"Our master plan is a diagram of what we want our homestead to look like when it grows up. It is a sketch that puts all the pieces together and helps us visualize the fulfillment of our dream and goals. It helps us in our decision-making and as we set our priorities. It serves as a visual reminder of what's been decided so far, because ideas fly and so does one's memory. By having a master plan, we can better determine how each step fits into the big picture."
"Developing a Master Plan,"
5 Acres & A Dream The Book *(p. 33)*

When *5 Acres & A Dream The Book* was published, it included our master plans from 2009 to 2013. We updated the plan again in 2014, 2015, and 2016. After that, there was a time when it seemed that our master plan was complete; that no more changes were necessary. However, a homestead is a work in progress. As our perception of our relationship to our homestead changed, our master plan had to adapt as well. This chapter shows you those adaptations in our master plans from 2014 until now.

Please note that the scale for all our master plans is approximate, as are the size and placement of buildings and fences.

2014

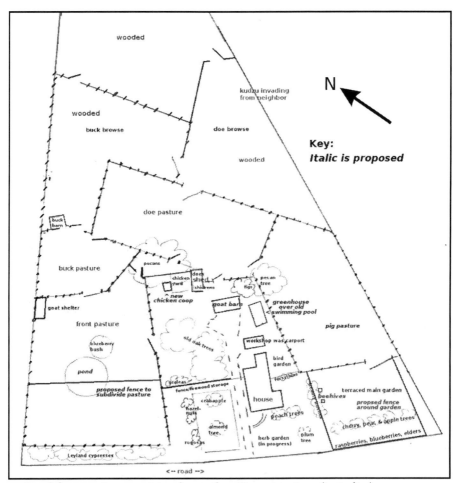

Road frontage is approximately 520 feet. Future projects (in italics) on our 2014 master plan were a greenhouse, beehives, pond, new goat barn, and pig pasture.

32 FINE-TUNING THE MASTER PLAN

Changes in 2014, included adding pigs to the homestead. Our breed of choice was the American Guinea Hog. This breed is small (adults weigh only about 200 pounds) and docile. They are excellent foragers. They were historically kept on small farmsteads because they fattened well without purchased feed, were prolific, and gave excellent flavored meat and plenty of lard. That sounded like the perfect pig for us.

> *"We brought him home in a pet carrier, the first of our two American Guinea Hogs. We let him out near the goat shelter in our front pasture, where he immediately attracted the attention of our bucks, chickens, and cats. None of the other critters knew what to make of him."*
>
> *"Pig Tales: Waldo,"* Critter Tales *(p. 296)*

Pigs readily consume garden and kitchen scraps, whey from cheesemaking, and waste from slaughtering and butchering.

We acquired an unrelated pair and named them Waldo and Polly. They were personable, loved to eat, and fit well into our homestead. They were excellent at turning food scraps, whey from cheesemaking, and butchering waste into meat for the table and manure for the compost. We no longer had to bury anything from home meat harvesting; the pigs relished it all, even the bones. We kept Waldo and Polly for breeding, but the piglets were for sale, barter, and meat.

One problem we addressed in 2014 was roaming dogs taking shortcuts through the garden to the barnyard where they would look for chickens and cats to chase. For that, Dan fenced the garden. Another problem was predation on chicks by rats, so we built a new and sturdier chicken coop.

"Even though I had done my homework regarding chickens and their needs, it was the day-to-day tending to them that helped to shape my idea of a better chicken coop. Ground breaking for the new coop was in February, and it took about three months of part-time building to complete. . . . It's roomier and brighter than the old chicken house, with easier access for both chickens and humans, plus it has storage space." "Chicken Tales: Moving Day for Chickens," *Critter Tales* (pp. 71-73)

2015

2015 was the year we planned our first forest garden hedgerow. It was to divide the front pasture into two paddocks. This plan also reflects our desire to expand the chicken yard and move the compost bins there. A location for honeybees was also decided upon.

The goat barn was the thing on which we couldn't decide. We'd come up with numerous designs over the years, but for some reason, building the barn never made it to the top of the to-do list. The greenhouse remained on the master plan, but because of our mild climate with its long growing season, it didn't seem as pressing as the barn.

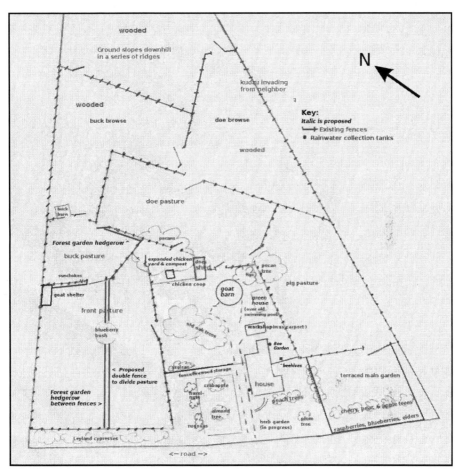

Our 2015 master plan with several proposed forest garden hedgerows.

2016

The goat barn finally became the priority project in 2016. It represented a huge undertaking because it meant more than just building a building. It meant rethinking fence lines, pastures, gates, traffic flow, and how we were using our fenced areas. This was the year we finally committed to a plan. You'll find photos of it on the following pages.

In May of that year, Dan purchased his sawmill, so that meant deciding on a place to set it up. To make working room for hauling logs and milling the lumber, we took down a few more fences too. During that time we had to shuffle critters around and the woods became off-limits to the livestock.

It was at this point we began to rethink the pigs. We got out our master plan to discuss and visualize options. Because of the fencing situation, we couldn't see any way around confining the pigs until lumber milling was done and the fences rebuilt. But we didn't want to do that. In the end, we decided it would be best to sell the pigs. I had just finished selling Polly's nine piglets, the freezer was full of pork, and if we could sell Waldo and Polly as a breeding pair, we could replace them later when everything was back in order.

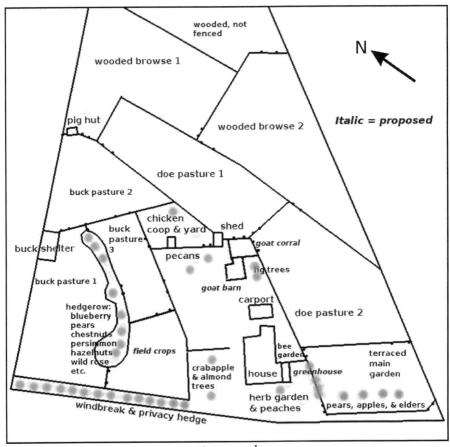

2016 master plan.

The 2016 master plan reflects progress on our first hedgerow, with its fence line following the contours of the land. Changes on the plan included the location of the proposed greenhouse and further subdividing the front pasture for an area to grow hay, grain, or other field crops. It also shows where we finally decided to build the goat barn and goat corral. After the barn was finished, the shed became Dan's workshop.

With building the barn, 2016 was a busy year for us. Following are the pictures I promised.

2016: Building the Goat Barn

Milling barn posts and beams started in May 2016. Curing time was one year.

Dan used a variety of large timber techniques in the barn's construction.

The tractor helped us raise the bents.

Left: The main part of the barn was framed out first, then the feed and milking room, and lastly the hayloft.

Below: An interior look at the metal roof. Dan began working on the roof in November, 2017. It was completed in January, 2018.

We didn't plan for a skylight in the milking room. We had previously purchased metal roofing panels for a project we eventually abandoned. They were two-feet too short, so Dan filled in the gap with translucent roofing panels to create a skylight.

After the roof, we installed and painted plywood siding.

Above: A view from the milking room. The head stalls for feeding make sure each goats gets their allotted ration (and theirs only!)

Left: The barn interior. A hay feeder is located directly beneath the hayloft hay chute. See "Resources" for where to find free plans for a single or double-sided goat or sheep feeder.

Right: Our ship's ladder to the hayloft. The gate at the bottom is to keep kids from climbing the ladder.

Below: Windows in the hayloft. These are entryway windows that we bought from a builder's surplus warehouse for $10 each. The loft has a solar shed light at the top of the ladder, but I appreciate the natural light during the day.

*Above: The hayloft filled with homegrown hay. The hay chute is in the center.
Below: Dropping hay down the chute is easier than carrying it across the barn.*

Finishing touches included a hand-painted barn quilt for the hayloft doors (above) and a cupola topped with a traditional barnyard weather vane (below).

The 2016 master plan remained unchanged for a number of years because all of our major permanent fixtures were established. These included fence lines, gardens, outbuildings, and fruit trees. With everything in its place, a visual map didn't seem as important. Any new ideas (such as the greenhouse) could fit into the landscape as we chose.

When I began work on this chapter, I realized I needed to show you our changes and new proposals since the 2016 master plan.

2019

For the 2019 update, I offer both an overview for a general lay of the land and a close-up on the next page for details.

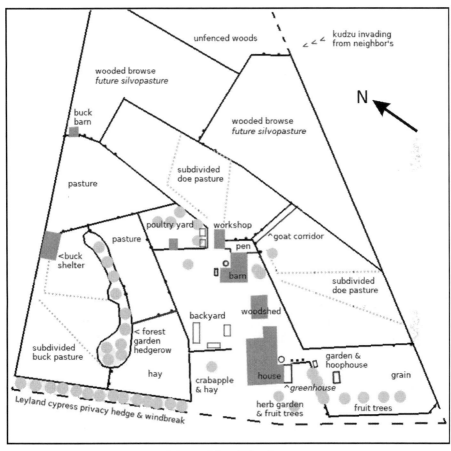

2019 Master Plan, The Overview.

Key: solid black lines = fence
  dashed black line = property line, not fenced
  pairs of black dots = gates
  dotted gray lines = electric fence to subdivide pastures
  gray circles = trees and bushes
  italics = planned projects

Forest garden hedgerow: blueberry, pears, mulberry, Japanese persimmon, chestnut, wild rose, pecan, oak, Jerusalem artichoke, chicory, echincea, acacia, plus grasses, ligustrum, sawbriar, and assorted weeds.

Fruit trees: apples, pears, cherry, and elderberry.

Woods: oaks, pines, maples, sweet gums, yellow poplars, dogwoods, wild cherry, eastern cedars, ligustrum, and magnolias.

FINE-TUNING THE MASTER PLAN

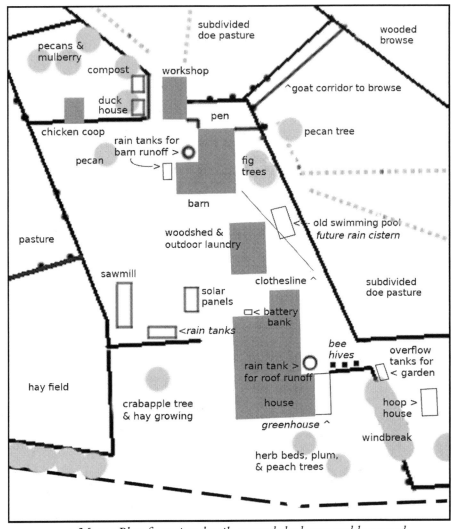

2019 Master Plan featuring details around the house and barnyard.

Between 2016 and 2019 we underwent a major shift in thinking as we learned about soil building through pasture rotation and cover cropping. I'll share more about that in upcoming chapters, but in terms of our master plan, we needed to figure out how and where to apply these concepts to our homestead.

For soil building, we realized that different methods would be needed for different areas. I'll discuss that in greater detail in chapter ten. For the master plan, we focused on ways to subdivide our goat pastures.

Subdividing the pastures for better grazing rotation was a challenge. It seems most proponents of intensive rotational grazing do so with meat

Our first gating system for paddock rotation.

animals that stay out in the field. Dairy animals are more challenging because they have to be brought in for milking twice a day. Then too, goats hate rain, and we wanted to provide shelter from both rain and bad winter weather. We had to devise a plan with laneways and gates to direct the goats where we wanted them to graze.

We used electric fencing for subdividing our pastures, and Dan built gateway arches using PEX pipe as conduits.

The girls quickly learned to respect the fence, but the bucks were harder to contain, and we had better success for them with electric netting. Either of these enables us to resize the paddocks easily based on forage growth.

While we worked on setting up the electric fences, the question of access to our woods came up. Dan had moved the sawmill down the hill and into the woods to mill closer to the source, but our new fence and gate setup offered no direct route for hauling freshly milled lumber up from the woods. We worked out a solution on the master plan. We shortened up the chicken yard by a tractor width and made a path for the tractor.

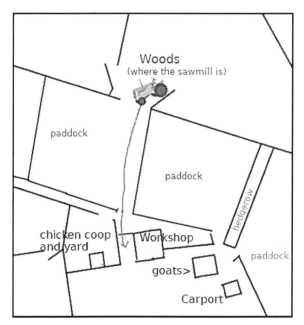

Left: The tractor path is a good example of how we use our master plan to discuss new ideas and potential changes. We tend to picture new ideas differently in our minds, so the plan offers us a common visual ground. We can run through any number of sketches on paper and once agreed, we make it a reality.

Below: chicken yard is on left, workshop on is right, tractor path and gate are in the middle.

Other additions to our 2020 master plan include a home for our Muscovy ducks and the solar panels. Also, potential locations for a future greenhouse and more rainwater collection tanks. Our revised version also

The gateway arches on page 47 work well for goat-size openings, but for the tractor path we needed to span a greater distance. The electric fence gate handles maintain the closed circuit needed for electric fencing, but make it easy to access the tractor path.

reflects our plan to establish silvopasture in our wooded areas. Silva is Latin for forest, so a silvopasture is a forest pasture. It's cultivated with both trees and forage for grazing animals. I've observed that our pasture forage does well where trees offer light shade from a high canopy of branches and leaves. As the old pines in our woods thin out and allow more light in, we will plant shade-tolerant species of forage to utilize new areas for grazing.

2020

Five and a half inches of rain one morning in February, 2020 caused us to reevaluate our master plan once again. It followed on the heels of an excessively rainy winter, and led to a new experience of weather extremes —flooding in the buck shelter.

Theoretically, we knew we should prepare for such extremes rather than averages. The problem is, we don't know what the extremes are until we experience them! Between October and January, we received over 28 inches of rain. The air temperature was seasonally cooler and the days cloudy, so the ground never dried out. When that early February rain came, our already saturated ground couldn't take any more.

Above: It was pouring rain when we headed out for morning chores, but the buck shelter was dry.

Below: Three hours later the shelter was flooded and the bucks were standing in six inches of water.

Left: As much as the boys hated standing in water, they were reluctant to wade through it to higher ground. It took the two of us to push and pull them to dryer quarters.

Fortunately, we had the old log buck barn to move the bucks to. Equally fortunate, the pasture had drained by the next morning. Of course, that experience leads to the question of avoiding a repeat performance in the future. How can we either prevent future flooding or otherwise facilitate draining?

One of the things we've learned about drainage problems is that they reflect infiltration problems in the soil. Healthy functional soil should be able to handle heavy rainfall without flooding. (See chapter 10 for more on soil). Creating fully functional soil takes time, however, so while we work in that direction, we discussed other options.

Ideally, we should look toward keeping as much moisture on our property as possible: dry wells, swales, or a seasonal pond, for example. When we drain the water off of the property, we drain topsoil, soil nutrients, mulch, humus, and organic matter from the property as well. One possibility was to resurrect the pond idea—not a permanent banked pond, but a round swale for directing seasonal flooding. Another possibility was to build a new barn for the bucks on higher ground in a more convenient location. This would solve other problems as well, such as the distance to carry feed, hay, and water.

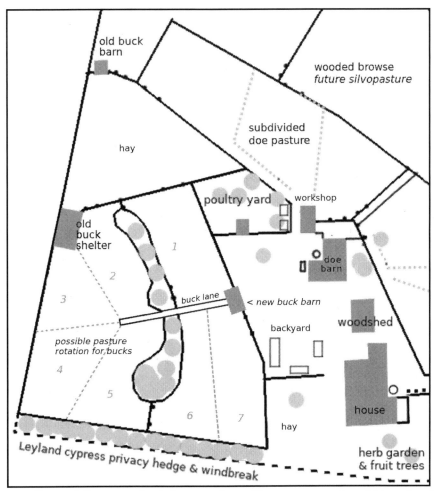

2020: Proposal for a new buck barn and possible pasture rotation. Using electric fencing (dotted gray lines) would make the number and sizes of paddocks flexible. Will this idea become a reality? You'll have to keep an eye on my blog to find out!

Looking to the future

As our idea for the buck barn and paddocks shows, analysis and discussion are ongoing. Another idea that hasn't made it to the master plan yet is to expand some our paddocks to incorporate more of our woods. That will require more fencing. We'd also like to have pigs again, which will require establishing a new area for them. Dan talks from time to time about getting meat rabbits, but has no concrete plans yet. As we brainstorm, we keep a copy of our master plan in hand. As a visual aid, it continues to be helpful, and I suspect it will be for many years to come.

CHAPTER 5

THE TRANSITION PHASE

"I like to call our beginning years of homesteading 'the establishment phase.' We have our land and the goal of becoming self-reliant, but it's going to take a lot get there: knowledge, equipment, tools, resources, and time. Because it is just the two of us, it is especially going to take time."
"The Establishment Phase," 5 Acres & A Dream The Book *(p. 47)*

As I sit at my computer and reflect on the five years since I wrote that statement, I find myself asking, "Well, are we established homesteaders yet? How would I describe our homesteading now?" As I try to figure out how to answer, I realize there is no clear-cut answer.

In our minds, establishing ourselves has meant infrastructure. It has meant building animal shelters and putting up fencing. It has meant creating vegetable and herb gardens, orchard, and pasture. It has meant adapting our living quarters to suit our intended lifestyle. It has meant acquiring appropriate tools and equipment. It has meant getting to the point when we can shift from big one-time projects to seasonal chores and maintaining what we've already done. So are we there yet?

If I were to evaluate our progress with a project checklist, I would say that we're at the tail end of our establishment phase. We have a few more windows in the house to replace along with exterior siding to put up and paint. We'd still like to build a greenhouse. But most of our outbuildings and fences are up, and our food-producing areas are coming along. We have much of the equipment and tools we need to reach our self-sufficiency goals, along with the knowledge and experience to use them.

What we've discovered, however, is that over time things change. This may be because something didn't work out as planned, or it may be based on knowledge and experience we've acquired along the way. A good example would be our chickens.

Our original plan for chickens was to let them free-range our pastures. Our chicken yard had two chicken gates. One allowed them into one pasture, the second into the other. This seemed like the best arrangement for happy, healthy chickens and so it was. The first problem presented itself when we created our first permaculture hedgerow to divide one of our pastures into two paddocks. We fenced the hedgerow with cattle panels, because we wanted to grow things in the hedgerow for our goats. The openings in the panels allow them to stick their heads through and enjoy the tasty forage growing there without demolishing it.

I planted the hedgerow with a variety of perennial edibles: pear, chestnut, mulberry, and persimmon trees, along with acacia and hazelnut bushes. I planted herbs such as comfrey, chicory, echinacea, thyme, and oregano. Everything was carefully mulched. However! The openings in the cattle panel are large enough for chickens to climb through. The chickens were drawn to that newly mulched hedgerow like bears to honey. It only takes a couple of chickens to unmulch everything in a matter of minutes. Even worse, underneath that mulch was fresh soil, loose from planting; perfect for digging up grubs and worms. The chickens not only scratched away the mulch and exposed the soil, but also scratched up the soil around newly planted trees and shrubs and exposed the roots. Not good.

So the chickens were banned from that pasture. Only the other yard gate was opened, so they could free-range on that side of the homestead. Imagine my surprise and frustration when I discovered that they had managed to find their way into the hedgerow anyway. In short order, our chickens re-scattered the new layer of mulch and dug up the freshly repacked soil. After several repeats of this not-so-funny comedy routine, they managed to kill most of my struggling treelings and bushes.

I ran into another chicken problem when I reseeded our pastures for forage improvement. Did you know that it only takes one or two chickens to eat a quarter acres-worth of pasture seed before you even discover they're there? I extended the height of the chicken yard fences by another two feet, and we began to look for alternatives to pasturing chickens.

I'll share more about those alternatives in chapter seven. The point here is how experience changes the way we address problems. In this case, it added new building projects to our list; projects we never anticipated when we drew up our first master plan. So because needs and plans change, a project checklist alone isn't a good way to evaluate progress. There are always new projects to add to the list.

Another way to evaluate our progress is by our lifestyle. Of that, I can tell you that our daily routine has changed. Infrastructure projects have become smaller, take less time, and require fewer resources. Such projects are now worked into the day, rather than working the day around the projects.

Further evidence of our lifestyle change is in the kinds of projects we now work on. In the beginning, some of our self-reliance goals—especially energy, and water—had to be delayed while we built and repaired. With more of our homestead foundation in place, we now focus on things like fine-tuning water filters for the rainwater tanks, experimenting with solar energy, and researching how to build a wood gasifier.

> *"Will we ever get out of the establishment phase and become truly self-reliant? I honestly don't know the answer to that."*
> *"The Establishment Phase,"*
> 5 Acres & A Dream The Book *(p. 56)*

So are we there yet? After ten years of homesteading, do we consider ourselves established? We've made progress on our projects and lifestyle, but at the same time have become aware of different challenges. The longer we homestead, the more I see it as part of a much bigger picture. It's more than having the constructed elements in place (garden beds, chicken coop, goat barn, solar panels, greenhouse, etc). It's more than the natural elements (weather, climate, growing conditions). It is also how we interact with all of these things because we're dealing with a whole of which Dan and I are just parts.

I used to think that we humans were in charge and that achieving our goal was simply a matter of careful planning, commitment, and willpower.

After all, our species is highly intelligent, so our role in the scheme of life is planning and implementation. Managing our resources means figuring out how to use them responsibly to meet our needs. However, as we celebrate our tenth anniversary on our homestead, I no longer believe that. Time and experience have shown me that nature's plan is already in place and that it is vastly different from how we humans think.

In nature's paradigm, the human role is to observe and respond. I am to discern the needs of my soil, plants, and livestock and meet their needs. If I focus on that, then our needs are met too. This is stewardship, but the problem is that it doesn't come naturally. To accept our proper role, one has to be willing to learn from nature. One must be humble enough to recognize mistakes for what they are usually, human assumption.

Learning about soil microorganisms is an example. I had heard of no-till from adamant adherents who proclaimed that tilling kills earthworms. Well, we've tilled and plowed enough ground to know that isn't exactly true. Tilling will not wipe out the earthworm population. I've done enough digging to know that numerous earthworms reside in soil too deep to plow or till. Because of that, the objections against tilling seemed ignorant to me. But when I started to research why our pastures weren't thriving, I learned that tillage isn't a problem for the earthworms, but for the microorganisms in the soil; the creatures too small to see with the physical eye—soil bacteria and mycorrhizal fungi. They are key to soil and plant health, and that's what tilling destroys.

Understanding that has meant more than changing the way we do things. It has also meant changing the way we see things. I used to say that we partnered with our livestock. Now I understand that it's more than that. We hold a key position in our homestead ecosystem as its stewards. Our responsibility is to serve it.

That has profoundly affected our goal of self-reliance. It's not Self that we should look to be reliant upon, rather, we need to be reliant on the natural world we've been given. The biggest challenge to that isn't gaining knowledge; it's extracting ourselves from the modern social system that demands our complete and total dependence upon itself.

Back to my original question: are we now established homesteaders? At this point, I would say we are in transition from establishment to stewardship. Because there are no clear markers along the way, there is no way to know how long this phase will take. Fortunately, it isn't a race and there isn't an expiration date. It's a process.

CHAPTER 6

FOOD SELF-SUFFICIENCY: FEEDING OURSELVES

"There are realities to food self-sufficiency that we didn't realize in the beginning."
        "Food Self-Sufficiency: Feeding Ourselves,"
          5 Acres & A Dream The Book *(p. 57)*

  Of our self-sufficiency goals, the most important has been producing our own food. It's also been the most challenging and time-consuming, but we approached it with a plan. We took care to choose a good site for a large main garden and prepare the soil. Then we planted fruit trees. The next year we added chickens and goats. Grain growing followed after that, and honeybees several years later. I knew part of the learning curve would include experimenting to find species and varieties best suited for our new location. But there was more to learn. In this chapter, I'll share the challenges we've faced and the adaptations we've made in the areas of food production, preservation, and storage.

Production

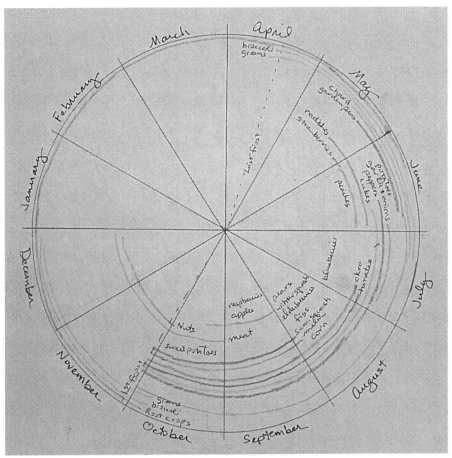

My harvest wheel, based on Masanobu Fukuoka's "Nature's Food Mandala."[†]

Our first garden was terraced and we tilled. During this time no-till gardening was becoming popular, with tilling beginning to carry an increasingly taboo status. But because of our wiregrass, I couldn't figure any other way around it. Wiregrass (also called "devil grass") is Bermuda grass (*Cynodon dactylon*). Bermuda is a very common summer lawn and pasture grass in my part of the country, because it is vigorous, tolerates drought, and holds up well under heavy traffic. But it is also extremely invasive and tenacious. It spreads by both stolon and seed and will quickly take over a beautifully mulched bed because it grows right up through it. I've pulled out stolons that measured nine feet in length before they broke off. That explains why mulch, no matter how thick, doesn't kill it. Even experts agree that wiregrass is impossible to eradicate.

Strawberry plant being strangled by wiregrasss stolons. I lost my first strawberry bed after it was choked out by wiregrass.

By the end of the summer harvest, the wiregrass in my garden beds would be so thick that it was impossible to hack through to the soil to plant again. So Dan would till and I would rake out as much as I could, in hopes of giving my next crop a head start before wiregrass swallowed it up.

I thought raised beds might help, but learned that it either grew up the sides and over the top of the bed, or it grew up from underneath. I ran into another problem too. During our hot, droughty summers my raised beds dried out quite quickly, no matter how well mulched. The hotter the air temperature, the faster this happened, so that they required more watering than my in-ground rows and beds.

Cabbages in a raised bed.

Climate change aside, part of our problem is that we live in the rain shadow of the Appalachian Mountains. Most of our summer weather systems come up from the Gulf of Mexico and travel in a northeasterly direction. Typically, they travel up the western side of the mountain range.

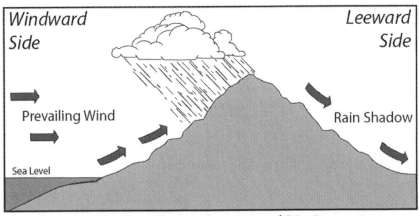

"Rain Shadow Effect" graphic courtesy of Meg Stewart,
[CC BY-SA 2.0 (https://creativecommons.org/licenses/by-sa/2.0)]

As the moisture-laden air rises to pass over the mountains it cools, condenses, and rains—on their western slopes. By the time the storm passes over the mountains and gets to us, it's rained out! This is called the orthographic or rain shadow effect and explains why it tends to be arid on the inland side of mountain ranges.

It's no fun to live with! Fortunately, we get enough rain (roughly 50 inches per year) so that we aren't in a desert. But we get enough hot, dry spells that it warrants paying attention to and planning for.

Since I wrote *5 Acres & A Dream The Book*, we have expanded our rainwater collection system. Of that, I'll share more in my chapter on water self-sufficiency. In terms of the garden, it was an excellent decision, because I now rarely have to ration water until it rains again. Even so, I knew that the better solution would be to retain more moisture in the soil for longer periods of time. We have worked toward that goal through soil building, swales, and an adaptation of hügelkultur.

Soil building is an important defense against dry conditions because some soils hold moisture better than others. Soil texture is an important contributing factor, with clay soils retaining moisture longer than sandy soils. But also, thriving microbiology is important. Mycorrhizal fungi in the soil produce a sticky substance called glomalin. Glomalin glues together soil particles, minerals, organic matter, and nutrients to form soil aggregates. Aggregates reduce water and wind erosion, reduce compaction, increase nutrient cycling, and increase water filtration and moisture retention around plant roots.[2] Look for more about this in chapter ten.

Swales are a kind of permaculture earthworks. These strategically dug trenches collect and hold rainwater. My first garden swale was a narrow trench and because it was too small to retain much water, it didn't help as I hoped. But I learned from it and made the next ones deeper and broader. I also combined them with the idea of hügelkultur.

Hügelkultur is a concept I learned about from Sepp Holzer's book, *Sepp Holzer's Permaculture*. The word is German for "hill (or mound) culture." These are constructed on the ground with dead logs, branches, grass clippings, cardboard, straw, manure, and other decomposable material. Layered and topped with soil, they become raised garden beds with enough organic matter to feed plants for a long time. Other gardeners were raving about hügelkultur beds, so I put some research time into the idea. While they seemed to work well in mild climates with fair rainfall, gardeners with hot summers and annual dry spells like mine were giving up on them. Hügelkultur beds appeared to work better in some climates than others. Mine was one of the others.

Research on our soil helped me understand the problem and find a solution. Through the online USDA Web Soil Survey (see "Resources"), I learned that our soil is classified as Cecil sandy loam. Our topsoil is a light brown sandy loam and the subsoil is classic southern red clay.

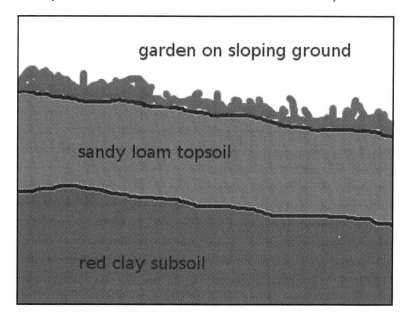

Our garden is on a slope, which means that water quickly drains through the sandy loam, hits the clay, and continues under the surface to

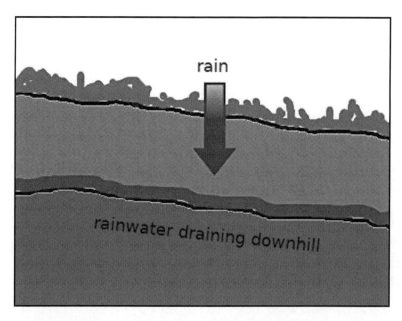

the bottom of the garden where it creates huge puddles. This is why the soil at the top of the garden always dries out more quickly than at the bottom. This information gave me an idea. Why not bury hügelkultur beds in swales? Here's what I've been doing.

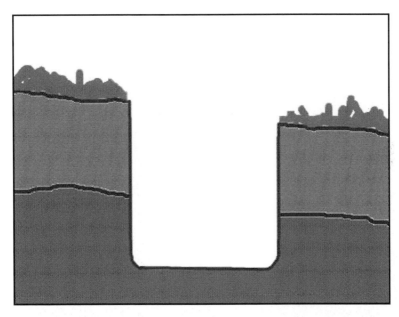

Bed-size trenches are dug into the clay subsoil and filled with rotting logs, branches, leaves, woodchips, compost, corn and okra stalks, and soil.

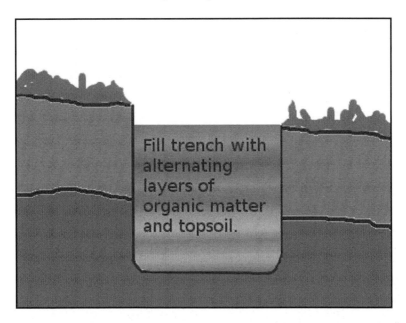

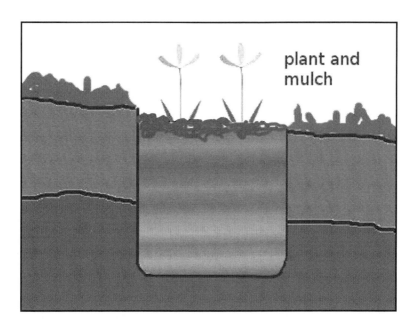

Now when it rains, the buried trench fills and retains rainwater.

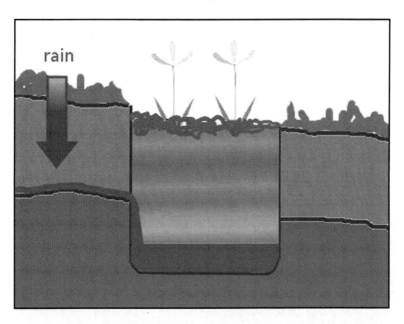

Between that and the increased organic matter in these beds, I water them less frequently than my other garden beds.

The finished beds are bordered. Note that they aren't raised beds, just bordered beds. I mulch the aisles with cardboard and woodchips to keep down the weeds. What I have found, is that with a good layer of mulch, these beds require less frequent weeding and watering. Because the beds are dug to a depth of about twenty to twenty-two inches, the soil is loose and full of organic matter. Tilling to remove wiregrass and make the soil workable is no longer required.

Gradually, I've been converting the entire garden to hügelkultur swale beds.

The healthier the soil is, the better it can retain moisture during dry spells. Our hügelkultur swale beds beautifully address the challenges of poor soil, hot dry weather, and even the wiregrass because the aisles are so heavily mulched.

Another challenge to learning to garden in our hot, often dry climate was figuring out what types and varieties of vegetables are best suited to our conditions. This challenge required several seasons of experimentation, with some successs and some failures. I had to give up on some things, and finally find the right variety for others.

> *"When it comes to food self-sufficiency, we really only have two choices. We must either learn to grow everything we want to eat, or we must learn to eat what we can grow."*
> *"Food Self-Sufficiency: Feeding Ourselves,"*
> 5 Acres & A Dream The Book *(p. 57)*

I can rarely harvest lettuce at the same time as tomatoes because of our climate. But I have learned which varieties work best. On the left, Lollo Bianda can take our winter worst. On the right, Jericho Romaine doesn't get bitter in our summer heat.

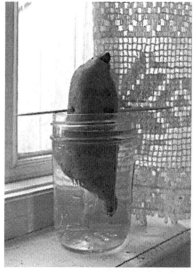

Middle left: Multiplier onions do better for me than globe onions. Plus, they are self-perpetuating.

Above: For me, Nancy Hall and Vardaman sweet potatoes do well.

Lower left: Foraged foods are also important, such as chickweed, a salad favorite.

One project with which I haven't been satisfied is our hoop house. The idea of a hoop house (also called a polytunnel) is along the lines of a greenhouse but without a source for heat. The clear polyurethane covering offers protection from frost and can extend the autumn growing season for many warm weather plants. It helps many cool weather crops thrive during winter as well. However, I've found that on mild winter days the hoop house can be extremely warm! Too warm for cool weather plants, which respond to the raised temps by going to seed. Another negative is that it prevents rain from watering the beds so watering becomes an extra chore.

Hoop house with polyurethane covering. On mild, sunny days the ends are open to help regulate the interior temperature.

In summer, the poly-cover can be replaced with shade cloth, to help things like salad greens, which benefit from protection from the blazing hot sun. Summer, however, is when I discovered how quickly raised beds dry out. With no insulation from the surrounding soil, the hoop house beds were always the first part of my garden to need watering. They required more frequent watering even though the beds were shaded by shade cloth. All in all, I haven't been convinced that the benefits, for me, outweigh the problems.

The garden is only one of our food growing areas. Our other food production areas include fruit producing trees and shrubs, a forest garden

Hoop house with shade cloth covering in summer.

hedgerow, herb garden, and an area for growing grain. Of grain, the challenge isn't growing it, but processing it for human consumption.

We started our grain growing with corn and soon added wheat. The experimental wheat patch pictured on page 61 of *5 Acres & A Dream The Book* did well, so we expanded it considerably. This past year I've been growing upland rice as an experiment. Upland rice can be grown without flooding. This makes it feasible when lowland, or paddy grown rice, isn't an option. I hope to grow rice in summer and wheat in winter, similar to Masanobu Fukuoka's natural farming method as described in *The One-Straw Revolution*.[3] He alternates these crops and uses the straw to mulch the next season's planting. He adds a little manure and grows clover as a companion crop. The result is building both soil and production.

Grain processing is a two-part job. The first part is threshing, i.e. removing the grains from the seed heads. Traditionally, this was done with a flail. A flail consists of a free-swinging pole attached to a wooden handle. When the ripe grain is beaten with the flail, the grains are knocked out. The second part is winnowing, which is separating the grains from the chaff (hulls and other plant waste).

Threshing is the most challenging part of growing grain. Others like us have invented various ways to accomplish the job: with a lawn mower, with a blender, with a bucket and chain, by converting a clothes dryer, even a baseball bat. An internet search for "threshing wheat" will yield all sorts of interesting ideas.

After trying several methods, Dan suggested we try our yard chipper. In *Critter Tales*, I showed you how he converted it to a feed processor for making goat feed.[4] We found it worked well for threshing our wheat.

Our little yard chipper works better as a wheat thresher than it did a chipper.

Eggs, milk, cheese, and meat are important as our sources of protein. Eggs and milk are seasonal foods because of their providers' production cycles. Meat is seasonal in the sense that the preferred time to harvest is autumn when the weather is chilly enough to not be pestered by flies or spoilage. As with garden produce, I've been working toward year-round production of eggs and milk.

For eggs, the simplest method for having fresh eggs all year is to keep a breed that lays during shorter daylight hours. We're far enough south that we can do that without artificially lighting the chicken coop. We've found Black Australorps to be a good breed for this.

For milk, we chose a breed of goat with the potential to give us a year-around supply of milk. Most dairy goats are receptive to breeding in the fall and have kids in the spring. Some breeds, such as Nigerian Dwarfs and Kinders, are considered aseasonal breeders. They can breed any time of year, which means they can kid at any time of the year. Ideally, staggering kidding would mean milk all year. I had to add "ideally," because such a breeding plan isn't as easy as it sounds. My Kinders have much stronger heats in late summer and fall, and subtle ones in spring. That means it's been more difficult to get those fall kiddings for winter milk.

I addition to goats, we keep Muscovy ducks for meat. Dan finds them easier to butcher than chickens, and we like the beef-like flavor of the meat. Muscovys make a good backyard variety of duck because they are quiet and rarely quack. They are perching ducks, which means they don't necessarily need a pond for swimming, although they dearly love to take baths. Dan keeps a small pool for them in the poultry yard.

Of other foods, my least successful category to date is sweeteners. Three potential sources for us are honeybees, sorghum, and sugar beets. You may wonder why I don't mention maple syrup, but it isn't produced here in the southern U.S., I suspect because of our warmer climate.

Of my sweetener options, I was most excited to get honeybees. *Critter Tales* tells the story of how I got my bees and why I chose Warré natural beekeeping.5 Here, I must continue where that tale left off and explain why we are now beeless.

My problem was skunks! Skunks are the number one varmint we catch in our live animal trap and rehome. Unfortunately, they are bee eaters. One popular deterrent is a skunk board placed in front of hive entrances. This is plywood or a scrap of carpet with nails poking up from the underside. Elevating the hive to where skunks can't reach them is another deterrent, as is fencing. But even with a skunk guard, all three hives of honeybees chose to leave anyway.

For several years, my poor hives stood empty. Then I got a copy of *Keeping Bees With a Smile* by Fedor Lazutin, a natural beekeeper who used a different style of beehive—the horizontal Layens hive. As much as I loved my Warré hives, I believe the Layens is a better choice for us, and so I plan to replace my old Warrés with Layens hives and try again.

One of my abandoned Warré hives with skunk guard.

Sorghum syrup was popular at one time in our part of the country, but it's something we've not tried to make. It would require a press to squeeze out the juice and vats for boiling it down to syrup. Acquiring equipment is always a big project for us, so the ability to make sorghum syrup still lies in the future.

A sweetener that I have made is sugar beet syrup. I found directions in *Grandpappy's Recipes for Hard Times* by Robert Wayne Atkins.[6] It was a lot of work to make the syrup, and so far, this has been a one-time experiment. I'm glad I tried it, but confess I prefer to use my sugar beets to feed the goats.

Another intriguing recipe from the same book is homemade yeast from hops.[7] I've been successful with sourdough, but also want to try

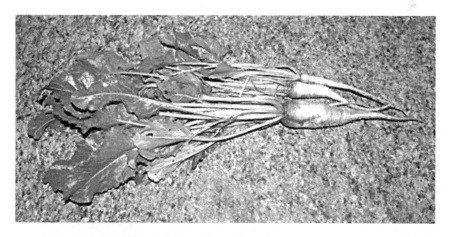

Above: Sugar beets. Most of mine reached a fair-sized growth, but in better soil they can grow larger. Both greens and chopped roots are relished by goats.

Left: Chopped sugar beets ready to simmer until tender. Then the pot liquor is drained off and cooked down to a thick and brown syrup. The cooked roots can be eaten (good with butter).

FOOD SELF-SUFFICIENCY: FEEDING OURSELVES

making homemade yeast from hops. My idea was to grow hop vines on a front porch trellis. It's an attractive vine and would shade the sunny west end of the porch. For three consecutive summers it made a promising start and then died. I'm not ready to give up on it yet, but next time I'll try a different location.

Preservation

Even though we're working toward year-round food production, I still preserve quite a bit. That hasn't changed, although I've given up on specific goals for food preservation.

> *"Initially, my method . . . was a pretty simple one. I considered how much of a particular food we eat each week, and then figured out how much we'd need on hand until next year's harvest."*
> *"Food Self-Sufficiency: Feeding Ourselves,"*
> 5 Acres & A Dream The Book *(p. 67)*

Now, we eat our fill of fresh foods and the remainder are preserved. I may end up with more than I need for the upcoming winter, but if the next summer's yield is poor, I'll have extra for the following winter too. Between that and expanding our fall and winter garden, we have the best variety we are able.

Reusable canning lids last for many seasons. See "Resources" for more information.

To my list of food preservation techniques—canning, dehydrating, freezing, and lacto-fermenting—I've added a new method, dry pack vacuum canning.

Vacuum canning is a way to preserve dry foods such as grains, dried beans, crackers, pasta, cereals, herbs, spices, flour, dehydrated foods, etc. By storing these foods in air-tight containers under a vacuum, they maintain freshness and remain bug-free. This idea appealed to me because of my two biggest challenges to food storage—humidity and pantry moths. I've tossed out more moth-riddled flour, grain, crackers, and dried fruit than you can shake a stick at. While the chickens may benefit from the moths' destruction, I certainly don't appreciate it!

There are a number of ways to create a vacuum in a filled jar, but I already had the perfect tool for the job. It's a "Pump-N-Seal Vacuum Sealer." I bought it for Y2K food storage. It uses plastic tabs as check valves in the lids of any jar. But by adding a jar sealer attachment made for the FoodSaver vacuum sealing appliance, I've been able to use regular canning jars and lids; no replacement tabs necessary. See "Resources" for where to find these items.

Dry pack vacuum "canning" with regular metal canning lids, FoodSaver attachments, and a Pump-N-Seal.

Experience has taught me that I can reuse metal canning lids that don't have bent rims. To test, I leave jars sealed with used lids on the countertop for several days. If I tap the lid and it bounces, the seal is broken and the vacuum is lost. I may retry or use a new, unused metal lid.

Top row: Canned black turtle beans on the left. Sliced cucumbers ready to dehydrate on right.

Middle row left: I put jars of leftovers in the freezer for winter soups. I add odds and ends of cooked veggies, meat, gravy, broth, cheese, rice, pasta, beans, etc.; anything except desserts! On the right is a gallon jar of lacto-fermenting veggies for kimchi.

Bottom: Vacuum packed beans, grains, & dried fruit.

Eggs, milk, and meat require different storage techniques.

Eggs. I've experimented with a variety of ways to preserve eggs: freezing, dehydrating, liming, and water glassing with sodium silicate (now sold as cement floor sealer). Even so, I rarely preserve eggs except in the refrigerator, unwashed to preserve their protective "bloom." This is because I've learned to adjust my meal planning according to the season. By keeping a winter laying breed I get enough for cooking and baking when daylight hours are short. My flock is small—one rooster and six hens —which has been adequate for eggs for just the two of us. Plus, I have less work in the preservation department and fewer hens to feed and house.

Left: Home-powdered eggs. Fresh eggs can be dehydrated either scrambled or raw and then powdered in the blender when crispy dry. It's time intensive, but the powder is easy to rehydrate and very handy! Dehydrated scrambled eggs are best powdered because the chunks are rubbery when rehydrated and cooked! For detailed instructions plus other egg preservation methods, see "Resources" for where to find my free eBook, How To Preserve Eggs.
Below: Preparing eggs for dyhydrating.

FOOD SELF-SUFFICIENCY: FEEDING OURSELVES

Milk. Our milk is preserved as cheese.

> *"Preserving milk through cheese making is an art form. Like many skills, there is a technical aspect, but there is also a knack to it that is developed with experience. My goal in setting out to learn this skill was not to make the traditional cheeses such as cheddar or Swiss. These require special starter cultures that I would always have to buy. Instead, I've experimented with readily available cultures, such as whey, yogurt, and buttermilk."*
>
> "Food Self-Sufficiency: Feeding Ourselves,"
> 5 Acres & A Dream The Book *(p. 71)*

Just between you and me, cheesemaking was a frustrating venture until I bought David Asher's *The Art of Natural Cheesemaking*. From it, I learned to use kefir as my starter culture. I also learned how to make cheeses more suited to my climate.

Most cheesemakers focus on northern European cheeses; cheeses that require several weeks of curing in temperatures between room and refrigerator temperatures with high humidity. This is how the cheeses with which we are familiar are made: Swiss, cheddar, gouda, for example. To make them, a cheese cave is necessary.

A cheese cave is a small refrigerator set between 45 and 58°F (7 and 14.4°C) with 80 to 98% humidity. These are the conditions necessary to properly age these kinds of cheeses. I've thought about getting a cheese cave, but honestly, don't have the room for it. That means I'm only able to make aged cheeses in spring and autumn when my pantry temperature is good for curing. For cheesemaking in summer, I started experimenting with Mediterranean and fresh cheeses.

Mozzarella (Italian) and Feta (Greek) are probably the most well-known Mediterranean cheeses. Less well-known cheeses such as Domiati (Egyptian) and Halloumi (Cypriot) are also good options. Traditionally, these kinds of cheeses are cured and stored in brine. Submersion in the salty solution prevents unwanted bacteria and molds from growing. The downside to brining is that the cheeses become progressively saltier. Some saltiness can be removed by washing the chunks in cool water. But my preferred storage method is to pack brine-cured cheeses in olive oil. This is an idea I got from *Preserving Food Without Freezing or Canning*. (See "Resources" for all the books I mention). Herbs, garlic, and peppercorns can be added to the oil to further flavor the cheese. After the cheese is consumed, the herbed olive oil is delicious on salads, for roasting vegetables, cooking eggs, or as dipping oil for chunks of fresh bread.

Homemade feta is stored in organic extra virgin olive oil with sprigs of fresh oregano, thyme, and rosemary.

Fresh cheeses can be eaten without aging. Paneer (Indian) or Queso Blanco (Mexican) are popular fresh cheeses. Another fairly common one is Farmers cheese, which is basically a cheese that hasn't been aged.

To preserve without a cheese cave, some of these cheeses can be frozen without affecting texture (the reason freezing isn't recommended for cheese). Mozzarella, Paneer, and Halloumi are cheeses that can be frozen. By making cheeses I can store in either olive oil or the freezer, I can have a year-round supply without waxing and without a cheese cave.

Whey is a byproduct of making cheese. When we had pigs, most of the whey was fed to them. Without them, whey quickly becomes a surplus item, because the chickens and cats are less enthusiastic about it. It has many uses, however. I use it for lacto-fermentation, to start new batches of sourdough, to replace water in soup, gravy, and bread, and for leavening baked goods.

"While doing my research for this project, I found quite a few not-so-common kitchen acids that can be used with baking soda to leaven baked goods. . . These substitutes include whey, brown sugar,

maple syrup, golden syrup, honey, tangy fruits, fruit juices, coffee, wine, pickle juice, sourdough starter, citric acid, beer or ale, sour cream, kefir..."

"*Not-So-Common Kitchen Substitutes for Cream of Tartar,*"
How To Bake Without Baking Powder *(p. 11)*

Whey can also be used as fertilizer and has shown promise as a deterrent for powdery mildew and other viruses in plants.[8] This is something I only recently learned but will experiment with in the future.

My favorite use of whey is to make whey cheeses. Milk contains two proteins—casein and albumin. Casein becomes the protein in a milk cheese, leaving the albumin in the whey. Ricotta is the most well-known albumin cheese, made by heating whey to near simmering so that the albumin separates from the liquid. An acid such as vinegar is often added to facilitate this process. The result is the mild soft white cheese we find in lasagna, cheesecake, and gelato.

A less well-known whey cheese is brunost, or Norwegian brown cheese. It is made by cooking down the whey until it becomes very thick. Then it is then poured into molds and allowed to cool. It has a delightfully nutty, tangy, sweet flavor; not at all like cheese. Dan and I prefer it soft and spreadable to eat with jelly on toast for breakfast. This version is called primost and is simply cooked for a shorter time than brunost. For long-term storage, I keep it in pint jars in the freezer.

To make brunost, whey is cooked down on the stove or in a crock-pot. To keep it from crystallizing, pour it into a glass bowl placed in ice water and stir until cooled.

Meat. My most common ways of preserving meat are by freezing or canning. While researching for *Preppers Livestock Handbook*, however, I discovered other off-grid meat preservation methods. Dehydrating, salting, brining, and smoking are all old-time ways of preserving meat. Two new-to-me methods were confit and mincemeat, which I have yet to try.

We've found that brining and smoking not only help in the preservation process, but also add flavor. My favorite way to finish brined and smoked meat is by slow cooking in either my crock-pot or my solar oven. That tenderizes it plus makes a smokey flavored broth. I use that for gravy, soup, or to can or freeze for smoke flavoring for future use.

Storage

Something that has turned out to be a bigger challenge than first anticipated is food storage. We don't need a lot of space to live in, but storing a year's worth of food, plus preservation equipment and supplies, requires space. When we initially evaluated our living space, we knew we needed something larger than the original 4-foot by 6-foot pantry. Instead, we converted a larger 8-foot by 12-foot room near the kitchen into a pantry. If we'd had a basement, we would have created storage space down there, including a root cellar.

One solution for lack of storage space has been to focus more on year-round gardening. My root vegetables store all winter in the ground with a good layer of mulch. Our winters are often mild enough so that I can grow hardy varieties of greens all winter. If the winter is more severe, however, we can't rely on the winter garden.

Summer presents storage challenges too. We live in the southeastern U.S. where summer highs are typically in the upper-90s (mid-30s). Because we don't use air-conditioning (more on that in chapter 8) my kitchen and pantry remain in the low-80s (upper-20s). The problem is that these temperatures affect the quality and nutrient content of stored foods. This can actually be measured with the Q_{10} temperature coefficient.

The Q_{10} temperature coefficient is a measure of the rate of change in a biological or chemical system as the temperature changes. It is measured using a change of 10° Celsius. For those of us who use the Fahrenheit scale, the change is measured per 18°F.[9]

The food industry uses "room temperature" or 72°F (22°C) as its basis for determining shelf life. Applying the Q_{10} factor, we can assume that shelf life is halved for every 18°F (10°C) above room temperature and doubled for every 18°F (10°C) below it. For nutritional longevity, all food items—fresh, canned, and dehydrated—should be stored at the coolest temperature possible.

Expiration dates aside, how long do canned and dried goods keep if the conditions are favorable? The only studies I've found are summarized at *Grandpappy's Official Website with Free Information on Many Topics*.[10] It cites two studies, one by the National Food Processors Association (NFPA), and the other by the U.S. Army. My summary of his summaries is that canned goods 40-, 46-, and 100-years old retained nutrients similar to newly canned foods.

Also cited on Grandpappy's website are two dehydrated foods studies by Brigham Young University. Their conclusions estimated shelf lives for many dried foods to be 30 or more years, when stored in airtight moisture-proof containers between 40°F to 70°F (4.4°C to 21°C).

What does that mean to me? That canned and dehydrated foods that are properly preserved and stored will keep for a very long time under the right conditions. In terms of my food storage goals, it makes sense to improve our storage conditions to obtain the longest shelf life we can.

All this got me thinking about how I use my two refrigerators and freezer and if there are alternatives. With goats and chickens, we have a large surplus of milk and eggs at certain times of the year. I can have up to ten half-gallon and quart jars of milk and eight dozen eggs in the fridge. But what else? I took an inventory and asked myself why I froze or refrigerated these items. My conclusion surprised me.

Most items were refrigerated to increase longevity. Dairy and meat are good examples, as are delicate vegetables such as fresh greens. Because our climate is hot and humid, I refrigerate anything that can get moldy quickly such as breads, figs, and berries, or opened jars of jelly, jam, and pizza sauce. Also, I refrigerate anything that can get soft and sprout in warm conditions, such as onions, garlic, and root vegetables. This is especially a problem in summer. After we get a string of days in the mid-90s°F (mid-30s°C), my kitchen and pantry temperatures gradually rise to about 85°F (29.4°C) during the day and drop to around 80°F (26.6°C) at night. When we get to about 100°F (38°C) in the shade outdoors, my pantry thermometer can reach up to 90°F (32°C). My kitchen is only a little cooler.

I also use my refrigerator and freezer to protect foods from insects. Pantry moths are my worst problem, but occasionally ants strike and then all sweeteners are moved to the fridge. To deter moths, most people recommend bay leaves but I've never found sources fresh enough to be effective for very long. Then too, considering the sizes and quantities I store of grains, flours, cereals, crackers, pastas, and dried fruits, it doesn't seem practical to keep fresh bay leaves in five or six dozen jars or more. I have found the vacuum sealing to be easier and more effective.

Without a root cellar or air-conditioned pantry, my fridge is used for long-term storage. I keep my season's worth of onions, garlic, potatoes, eggs, lacto-fermented vegetables, and rendered fat in the fridge. I also use it to store garden seeds, bulk grains, essential oils, plus livestock vaccines and antibiotics.

My freezer preserves meat, some cheeses, cream, colostrum, rennet, berries, nuts, chunks of melon, pureed winter squash, and homemade ice cream. I also freeze small amounts of tomatoes, berries, figs, and bones for bone broth. By

In the fridge, I decrease my use of plastic by storing leftovers in glass jars and bowls with glass saucers for lids.

cooler weather, I have enough for a canning session, with the added benefit of canning after the summer heat has passed. Homemade convenience foods are stored in the freezer as well, particularly unbaked pies and jars of leftovers for wintertime soups.

After making and categorizing this inventory, I asked what could be changed? Giving up air conditioning was noble on the one hand, but created problems on the other. In looking over my lists I realized how few of those items actually need refrigeration. Many of them simply need cooler storage temperatures than my pantry and kitchen provided. Dan and I discussed our options and came up with a plan, along with the steps it would take to accomplish it.

The first step was to move the chest freezer and my back-up fridge out of the pantry. We moved the freezer to the back porch and put it on its own solar-powered energy system. Then, we replaced the old fridge with a low energy chest model. I'll tell you about all that in chapter eight.

The benefit of moving the appliances was to free up room in the pantry for more shelves. Also, we eliminated a source of heat, i.e., the compressors of these appliances.

Freezer and new fridge, a small converted chest freezer.

Future improvements planned for the pantry include new energy-efficient windows and better insulation in the walls. Dan also wants to experiment with low energy cooling through buried ductwork and improved ventilation with a solar attic fan.

Another idea we continue to contemplate is a root cellar. We discuss it from time to time, but we've never finalized a plan because unanswered questions remain. How cool would a root cellar be in our climate? Would it be enough to make the work of building it worth it?

A Year's Worth of Food

If our goal is food self-sufficiency, then the last question is likely how well we're doing. How satisfactory one would find the answer to that, would likely depend on what's considered an acceptable diet. We're able to produce and preserve food for year-round eating, but our diet is still subject to our regional climate and whims of the weather. Those pretty much determine how much diversity our homegrown diet has.

Our diet is seasonal and meals focus on whatever is producing well at the time. This doesn't mean I never shop at a grocery store, but my shopping list is pretty basic. It sticks to staples we can't or don't produce ourselves such as olive oil, sea salt, and unbleached flour, a few particular favorites like black olives and bananas, and sometimes items to fill in a nutrient gap, such as carrots for vitamin A. I also take advantage of good deals for stocking up. This kind of menu planning keeps my grocery bill down, but it has a downside too. Preparing meals from the season's abundance means day after day of pretty much the same thing to eat. When our chickens are laying well, we eat eggs every day, usually for lunch. But how many ways are there to eat eggs? A lot, actually, but before the summer is over I still get tired of eating them and find myself hankering for a tuna fish sandwich!

Transitioning to a homegrown diet has meant finding substitutions for many familiar food items such as condiments. For example, mayonnaise on sandwiches. I've tried substitutes, but nothing comes anywhere near it. Yes, I could make my own, but I would still have to purchase the oil to make it. While I may feel good about using my homemade mayo on sandwiches and salads, am I really any closer to true food self-sufficiency? Add to that the time and need for a noisy high-speed blender, and I have to confess that I'd rather just buy it. Shocked? These are the kinds of trade-offs one makes when trying to balance time and food preferences with a transition to food self-reliance. If it comes down to it, I can use butter on my sandwich and kefir as salad dressing.

While I've learned to substitute a lot of ingredients, we've found that certain foods just aren't the same. Pizza, for example, just isn't "right" with whole wheat crust. The same is true for French bread. It doesn't seem like French bread unless it's made with white flour. I admit these are food prejudices in which we indulge ourselves. But on the other hand, the goal isn't to achieve a type of food legalism; it's making the best with what we've got while maintaining some sense of diet normalcy.

Addressing the challenges of production, preservation, and storage has significantly shaped our lifestyle. They are the foundation of our seasonal cycle and continue to influence changes we consider making.

Notes

[1] Masanobu Fukuoka, *The One-Straw Revolution* (NYRB, New York, 1978) 131.
[2] Kris Nichols, "Does Glomalin Hold Your Farm Together?" *USDA-ARS-Northern Great Plains*, https://www.ars.usda.gov/ARSUserFiles/.
[3] Fukuoka, 1.
[4] Leigh Tate, *Critter Tales* (Kikobian Books, 2015) 189-190.
[5] Ibid, 317.
[6] Robert Wayne Atkins, *Grandpappy's Recipes for Hard Times* (Grandpappy, Inc., 2011) 9.
[7] Ibid, 6.
[8] Gillian Ferguson, "Milk as a Management Tool for Virus Diseases," Greenhouse Grower Notes, *Ontario Ministry of Agriculture, Food and Rural Affairs*, Nov. 1, 2005, accessed Nov. 11, 2019, http://www.omafra.gov.on.ca/english/crops/hort/news/grower/2005/11gno5a1.htm.
[9] Ken Jorgustin. "Temperature Versus Food Storage Shelf Life," *Modern Survival Blog*, January 5, 2015, accessed August 15, 2019, https://modernsurvivalblog.com/survival-kitchen/temperature-versus-food-storage-shelf-life/.
[10] Robert Wayne Atkins, "Five Different Shelf Life Studies: Two on Canned Food and Three on Dry Food," *Grandpappy's Official Website*, 2007/2010, accessed April 11, 2020, https://grandpappy.org/hshelffo.htm.

CHAPTER 7

FOOD SELF-SUFFICIENCY: FEEDING OUR ANIMALS

"Animals are an important part of our homestead. . . It has been important to consider how they fit into our overall goal of self-sufficiency. We decided, in the beginning, that each animal must contribute to our needs, and that we, in turn, must not keep more than our land can properly provide for."
"Food Self-Sufficiency: Feeding Our Animals,"
5 Acres & A Dream The Book *(p. 83)*

Feeding our animals from the land has been the most challenging and elusive of our self-sufficiency goals. The reason is a good example of one of the problems many new homesteaders face as they try to sort out how to approach this lifestyle. This is one of those "I wish I knew then what I know now" goals because it would have saved us a lot of frustration along the way, not to mention time.

So what was the problem? The problem was that I approached critter feeding from the wrong philosophy—that of the industrialized production model. It wasn't that I deliberately chose this model. It was that I had no experience and the first books I found on the subject presented this model as the standard way to do things. To be self-sufficient, I thought it was simply a matter of imitating the commercially formulated packaged feeds by substituting homegrown ingredients.

Considering my level of knowledge and experience, that seemed like a good goal at the time. Had I understood there are other philosophies for feeding livestock, I would have saved myself a lot of time and trouble! The information was out there, but coming across it represents the journey of discovery that so many of us face. Looking back now, I can say that the learning curve has been long and steep, but from it, I've learned important lessons.

In our early years, I focused on research and experimentation with the assumption that goats, like cattle, need concentrates. That's what the books said. That's what the feed experts said. That's what the hobbyists said. Concentrates is a generic term for feed such as grain or, more commonly, the packaged kibble that most of us buy for our critters. These are convenient because they contain the recommended amounts of nutrients. The objection is that the main ingredients are commercially grown corn and soy, both of which are genetically modified. I thought I'd have to grow my own. The bigger problem, however, was that feed bag labels list ingredients as "products," such as roughage products, forage products, and grain by-products. How was I supposed to know what to grow from that?

For a number of years I worked toward growing and mixing my grains and legumes as a replacement for packaged feeds. While the weeds thrived, crop yields were less than spectacular. I was able to grow enough wheat, corn, and cowpeas for Dan and me, but never enough to feed all our critters too. Not enough until the next harvest, anyway. This was a frustrating problem for me.

The epiphany came when I was writing *Critter Tales* and researching growing our own goat feed.

> *"What does all this mean to me in terms of growing grains and legumes for our goats? It means that if I can provide high-quality forage, both as fresh pasture and browse plus dry as hay, then I do not need to focus on growing grains and legumes for them."*
> "Goat Tales: Toward Sustainable Goat Keeping,"
> Critter Tales *(pp. 182-183)*

Upper left: Ozark Razorback cowpeas, a small variety that chickens can eat. Goats will eat the whole pods, plus leaves and vines.
Upper right: Jerusalem artichokes, washed and ready to chop for the goats.
Lower left: Amaranth heads drying. I toss whole heads to the chickens. Goats can eat the seeds, leaves, and stalks if chopped.
Lower right: Cushaw winter squash. Like all winter squash (including pumpkins) chickens and goats, both, can eat the seeds, flesh, and rind.

Left: I dry herbs and greens on a wire shelf between layers of fiberglass window screening. These become my homegrown vitamins and minerals for goats.
Right: Sprouting grain for all species adds nutrients. It also adds bulk and can cut the feed bill by up to half.

I began to understand that modern pelleted feeds represent a philosophy of livestock feeding. This philosophy is geared toward the industrialized model which optimizes protein for maximum production of eggs, milk, and meat. As a novice homesteader, I bought into this because, of course, I wanted my critters to "reach their genetic potential." I mean, who wants a puny chicken when one can have a nice, big, fat one?

There are problems with this model, however, that gradually became obvious to me. It finally occurred to me that "genetic potential" is a misleading term. For example, my dad is six foot one, so I have the genetic potential to pass that on to my children. That doesn't mean my children will grow to be as tall as my dad, but based on the genes I contribute it's a possibility. However, I don't control that. I can feed them super-powered foods, but other factors determine how that genetic potential fleshes out.

The same is true with animals. Feeding my goats a high-protein, high-carbohydrate feed won't automatically turn them into two-gallon per day milkers if their DNA dictates otherwise. In fact, it can be detrimental because their digestive system, the rumen, isn't designed for highly concentrated foods such as grains and kibble. It is designed to slowly extract and convert the building blocks of protein and carbohydrates into what they need. And where are these building blocks found? In good quality forage and hay.

The other problem is that the modern production model is trying to keep as many animals as possible in as small a space as possible. That completely changes the playing field, because with pasture our critters can feed themselves. They don't need complete nutrition in one package.

Understanding these things brought about my first change in thinking. And doing. I didn't immediately stop feeding a concentrated ration, but I began to think of it as a supplement rather than the main course. The main course should be forage and hay with top nutritional quality. I began to focus on growing better pasture.

We were already working on correcting mineral deficiencies in our soil. I detailed this in *5 Acres & A Dream The Book*. Then I planted pasture grasses, legumes, herbs, even root crops to feed the goats and improve the soil. For winter grazing, I planted a deer forage mix of wheat, oats, and Austrian winter peas, plus annual rye. Satisfied that we'd done a good job, I assumed that was it.

Unfortunately, that wasn't it. We had good pasture for the first season or two, but it gradually deteriorated and grew more weeds than desirable forage. In part, I can blame the weather because we tend to have long droughty heat waves every summer. In an attempt for a wide diversity, some of what I planted wasn't drought-resistant. But also, I had to blame my ignorance because we knew so little about pasture maintenance.

A polyculture of wheat, oats, winter peas, clover, and vetch for forage or hay.

As with feed regimens, there are different philosophies about growing and maintaining pasture. The commercial production model is easy-care monoculture. In my part of the country, this means tall fescue for winter pasture and Bermuda grass for summer. Both are popular because they are hardy, drought-resistant, can handle heavy traffic and grazing, and are able to choke out other species to create a commercially desirable single-species pasture. But tall fescue has problems with endophytes, and Bermuda is aggressive and invasive. This is the "wiregrass" that becomes a weed in the garden. It dominates and frustrates efforts toward forage diversity.

For my purposes, the more desirable pasture is a highly diverse polyculture. This is the model we see in nature but achieving it was another matter. In researching sustainable pastures I discovered several things. One was that even perennial pasture species only last for three to five years. The other problem is the livestock themselves, who will eat down what they favor until it cannot perpetuate itself. That leaves room for things they don't eat to thrive. This pointed to the need for rotational grazing. My first rotation plan was to divide the pasture in half.

As more and more bare spots appeared, I turned to annual seed mixes of sorgham-sudangrass and clover for summer and wheat, oats, and peas for winter. These are less expensive than perennials and more readily available. Theoretically, annuals should work if one can coordinate planting according to their seasonal cycles. If timing and conditions are right, then livestock should have uninterrupted year-round grazing. But somehow it never worked out that way, mostly because of when the rain came. So in between seasons, we had scanty growth, poor grazing, and long stretches with little for the goats to eat.

About this time, I read *The One-Straw Revolution* by Masanobu Fukuoka. His natural farming method of growing grain is brilliant, yet simple. He doesn't plow, till, or add fertilizers. In summer, he grows rice; in winter he grows a mix of barley and rye. After threshing, he uses the straw to mulch the growing plants, tosses on a bit of chicken manure to help decompose it, and that's about it.[1] This ongoing cycle is not labor-intensive, yet productive. I loved the book but didn't think it applied to me because our homegrown wheat straw is valuable as hay or bedding.

I began to connect some dots when it came time to clean out the goat barn. I use the deep litter method for bedding, so the goat barn contains a gold mine of straw, wasted hay, manure, urine, and barn lime. This kind of lime—also known as pickling lime or calcium carbonate—helps with odor and insect control. It is just alkaline enough to deter insect eggs and larvae. It's also used as agricultural lime to adjust pH in acidic soils.

Instead of using the soiled bedding for compost, I decided to use it directly on the pasture in Fukuoka fashion. My pasture seed mix consisted of locally bought annual seed and mail order perennial pasture seed. I broadcast it anywhere bare soil showed and then spread a fluffy layer of barn litter over the seed. The barn litter contains enough manure to help decompose the straw. An unexpected plus is that because of the manure and urine, the goats leave these areas alone instead of trying to eat any grain seeds on the ground. I found that they pretty much ignored the new growth until it was about four or five inches tall. With my modified Fukuoka planting, I didn't have to banish the goats while the seed grew. I was delighted when the method worked very well.

Even so, I didn't get the established perennial pasture I was hoping for. Pasture grasses, legumes, and edible weeds made a good start, but gave way to woody annuals, horse nettle, and ground ivy; all things the goats won't touch. I knew our regional soils are poor but the mineral amendments didn't seem to change what grew in them. The perennial grasses I did manage to grow would die back during long dry spells. When I asked what perennial pasture seed my favorite feed and seed store recommended, she just shook her head. Commercial perennial pasture grasses don't thrive in our area, so most farmers planted annuals. The worse the problem got, the

My modified Fukuoka method of planting pasture consists of seeding bare areas and then covering the seed with old barn bedding as mulch. I have much better germination and healthier growth with this method.

more discouraged I became. The poorer our pasture, the more feed I had to buy. I had a goal I couldn't seem to reach and couldn't figure out why.

The missing puzzle piece came when I read Joel Salatin's *Salad Bar Beef*. There it was: poor forage points to poor management. The answer is intensive rotational grazing. This meant more than the two paddocks I occasionally switched out. As I read, I began to understand the plant growth cycle and how grazing animals impact it. I learned that the first step toward weed control is serious rotational grazing. By concentrating stock in a relatively small area for only a day or so (mob grazing), the edible forage has time to recover and thrive. Joel says the forage in a grazing area can be completely changed with this management method.[2]

The pioneer of this technique is Allan Savory. In his book, *Holistic Management: A Commonsense Revolution to Restore Our Environment*, he explains. When livestock are concentrated in small areas they either eat the forage or trample it. The key is to move them after they've eaten the top half of the plants and no more. With the help of any manure left behind, the trampled forage begins to decay which in turn feeds soil organisms, retains soil moisture, sequesters carbon, and builds the soil. The long rest period allows forage to recover fully before being grazed again.[3]

The goats mob up as they compete for the best bites of fresh forage. When rainfall is adequate, the system works well. When we hit dry spells, forage becomes scant.

Sounds crazy, I know, because the high stocking density these men recommend pretty much goes against everything we've assumed about livestock and grazing. But when managed properly, the results are phenomenal, as reported by everyone who practices it. Allan Savory, for example, is using this method to turn desert in Zimbabwe back into grassland. Missouri cattle rancher Greg Judy has been able to stop buying pasture seed, fertilizer, and hay, eliminate the use of machinery, yet has doubled beef production and increased forage quality and diversity.[4]

Considering the problems I was having with our pastures, I was ready to try anything. I started with my two doe pastures, and we used electric fences to subdivide them into smaller paddocks (See 2019 master plan, page 45). Then I started rotating where the goats grazed. As I watched them in the paddock, I began to understand what Joel means by mob grazing. Because the goats are in a relatively small area, they are competing for the best bites and aren't so choosy. They form and move as a "mob." I keep an eye on how much they eat and move them to another paddock after they've eaten off the tops of all the available forage.

Trimmings from bushes, trees, grasses, and weeds make for good goat goodies. Putting them in the hay feeder protects them getting trampled on the ground.

Right: Cattle panel fencing allows our goats to have restricted access to areas such as our forest garden hedgerow. They are able to eat but not demolish the plants that grow there.

Below: Goats tend to be greedy about feed, with the biggest and strongest pushing the others away. To make sure every goat gets their fair share, I separate them according to age group. Younger goats are fed under the loafing overhang, while adults are tied in individual head stalls.

Forage is seasonal, so folks who utilize this management method successfully, acknowledge that it takes years to see a full transformation of their grazing areas. Dan and I implemented it the same year I started writing this book, so my readers will have to wait until the next sequel or follow my blog to find out how well we're doing with it! We plan to continue rotating where the goats graze and continue seeding bare areas with my modified Fukuoka pasture planting. To that, I've added sprinkling the bare soil with minute amounts of the minerals our soil is deficient of. For the rest, I trust nature to do what it was designed to do, especially now that I'm cooperating with it.

One last consideration of the grass-fed model is the animals themselves. Like people, individuals have different energy and nutrient requirements. I can feed two goats the same amount and one will get fat while the other will lose weight. Producers who have gone to grass-fed meat and dairy systems report that there is a selection process. Animals that can't thrive on grass and hay only are culled, either by nature or by human selection. Eventually, the herd is one that is able to thrive on forage and hay rather than large amounts of concentrates.

For now, I supplement my goats' foraging and hay with homestead-grown feed and add as much purchased grain as necessary to keep them in healthy condition. Each year, I sell enough registered Kinder goats to cover the cost of feed, hay, and supplies. That means our goats are self-supporting which is the next best thing to being self-sustaining. As our pastures and hay improve in quality, I will be able to decrease the amount of feed and hay that I need to buy.

What about our chickens and ducks? When they free-ranged, they were able to forage much of their food. When we began seeding and mulching the paddocks, however, they became counterproductive. Both love grain and will readily scratch and shovel through my newly planted and mulched areas, eating all my seed and seedlings! That problem led to a discussion about a tough decision. Were we going to continue free-ranging them? Or were we going to consider confining them? Over the years we've expanded the poultry yard but would it be enough space if the birds no longer had access to pasture? If they couldn't range the pasture, what could we do to make up for it?

The first step was to ask ourselves how many chickens and ducks we really need. If their ranging area was going to be smaller, shouldn't we consider keeping fewer? We had to ask ourselves how many eggs per week do we need? How much meat?

Some people warn against keeping chickens and ducks together, but we've never been able to keep them apart. We offered the ducks their own space, but they seem to think they have a right to whatever the chickens have. The chickens disagree, so there are numerous squabbles over chicken scratch and treats. The rooster and drake defend their ladies and make sure every bird keeps their distance.

> "You can begin to see the logistical problems with this: having enough eggs to successfully hatch that many, then feeding, housing, raising, and processing all those birds. . . . Then too, more chickens mean raising and storing more feed, along with larger housing facilities and the job of keeping those facilities clean."
>
> "Of Chickens, Goals, and What I've Learned," Critter Tales *(p. 91)*

Of chickens, we settled on six hens and a rooster. That gives us enough eggs to meet our needs plus an occasional chicken dinner. For meat, we keep Muscovy ducks. Muscovy is unlike other duck meat—less fatty and more beef-like in flavor than poultry. Plus, Muscovies are exceedingly prolific. They are tremendously broody and excellent mothers, able to raise more than one batch of ducklings each summer. One drake and one or two ducks can keep us in quite a bit of meat.

What else could we do? We implemented several new ideas. The first was to move the compost bins into the chicken yard.

The chickens are more than willing to help with the compost and come running when they see Dan or me arriving from the house with the compost pail. They are surface scratchers, so we assist by occasionally turning the piles.

> "Chickens are given free access to huge compost piles containing barn cleanings and food scraps . . . They constantly dig through the pile, feasting on the scraps, insects, earthworms, and whatever else they can find. The appeal of this to me was how it partnered with the chickens, utilizing their natural behaviors to feed them while producing a valuable product in the process. It is a work-smarter-not-harder method for needful things on the homestead."
> "Of Chickens, Goals, and What I've Learned," Critter Tales *(p. 88)*

I spent years trying to keep chickens out of the compost. Now I realize that was a big mistake! Not only do they find food scraps, grubs, and worms to eat in the compost, they also speed up the composting process. I don't know exactly how, but there is a symbiotic relationship between chickens and compost! On top of that, it makes composting easier. Chickens are surface scratchers so every couple of days or so the piles still need a deeper turning. This formerly dreaded job is now fun because it makes the chickens so happy. Turning the pile stirs up new tidbits which are quickly snatched up with delight.

The second thing we did was to construct grazing beds. These are frames of any size, topped with wire fencing. We used welded wire fencing with 2-inch by 4-inch openings, which is strong enough to support the weight of several chickens. Seed is planted in the bed, and after it grows

Chickens working a grazing bed.

through the wire the chickens graze the greens. The ducks are grazers too and are happy with the fresh greens as well. Eventually, it doesn't grow back, so then we move the frame for a new planting. The chickens scratch through the reamins in search of anything good to eat.

How well has all of this worked? The best answer for that is the amount of feed the chickens now consume. We keep a filled feeder in the coop which gets pecked at only occasionally, but it takes months to finish a 50-pound bag of feed. We've cut back considerably on scratch too. Dan still tosses them a couple of handfuls every morning and evening, which satisfies chickens, ducks, songbirds, and squirrels. Something else that helps is that our large poultry yard is completely shaded by mature pecan trees. These provide some protection against hawk attack plus drop a lot of leaves that decay and make black soil. Earthworms abound and keep the chickens busy hunting for them. From a diet of kitchen and garden scraps, fresh grass, grubs, grain, and earthworms, our flock is happy and healthy: alert, bright-eyed, and with sleek shiny feathers.

Before I close out this chapter, I'd also like to mention pigs. We finally got our pigs in 2014, and for several years we kept a breeding pair of American Guinea Hogs. This is a small, friendly breed that gave us plenty of piglets for selling, trading, and meat. They were easy to feed because they were good at foraging. They happily consumed any surplus of milk, whey, eggs, produce, kitchen and canning scraps, plus butchering offal, including bones and skins. I've written about their stories and antics in *Critter Tales*, which was published before we began to have problems.

The problems weren't with the pigs, however, the problems were with our trees and fences. I'll address this in detail in chapter ten, "Resource Self-Sufficiency," but the nutshell version is that we began to have trees falling in our woodlot and destroying fences. The pigs spent quite a bit of time in the woods, and the prospects of escape or being hit by a falling tree were concerns. The only option was to confine them, which we didn't want to do. So we sold Waldo and Polly with a promise to ourselves that once we were set up for them we'd get pigs again.

A Year's Worth of Feed

The goal of feed self-sufficiency for our livestock remains a priority, but the plan to achieve it has changed. Through experimentation and research, we have a new way of understanding the relationship of our animals, our land, and our role as stewards of both. Grazing animals are a natural part of all grassland ecosystems, so rather than trying to grow and stock up on a year's worth of feed, our focus now is on developing a forage-based feeding system. In addition to pasture, we work toward growing top-quality hay, including unthreshed small grain grasses such as wheat and oats. For nutritional support, I add greens and root crops, plus my dried herbal mineral mix.[5] This is how livestock thrived before the advent of packaged kibble, and this is how natural farmers are raising thriving livestock again. I wish it hadn't taken us so long to understand all of this, but at least we're on the right track now.

Notes

1. Masanobu Fukuoka, *The One-Straw Revolution* (New York: NYRB, 1978) 1-3.
2. Joel Salatin, *Salad Bar Beef* (Swope, Virginia: Polyface, Inc., 1995) 46.
3. Allan Savory & Jody Butterfield, *Holistic Management: A Commonsense Revolution to Restore Our Environment* (Island Press, 2016) 57-58.
4. Boyd Kidwell, "Mob Grazing: High-density stocking builds profits back into into the cattle business," *Angus Beef Bulletin*, March 2010.
5. Leigh Tate, *5 Acres & A Dream The Book* (Kikobian Books, 2013) 219.

Chapter 8

Energy Self-Sufficiency

"I confess that energy self-sufficiency is a goal that, at this time, seems beyond our reach. . . . That does not mean, however, that we give up on the goal. It means that we accept the reality of our situation and take whatever steps we can."
"Energy Self-Sufficiency," 5 Acres & A Dream The Book *(p. 122)*

Energy self-sufficiency seemed an elusive goal when that was written almost six years ago. It still does because it seems unlikely we'll ever be able to completely abandon the electrical grid. On the other hand, we have gradually taken steps to be less dependent on it. How successful have we been so far? Two things that show progress are our monthly electric bill and changes in our lifestyle.

We used to have huge seasonal fluctuations in our electric bill. Our house is all-electric, which means the stove, water heater, and heater in our HVAC unit are all electric. Electric heat is a huge energy hog. When we relied on it, our electric bill could be as much as $200 per month.

That really put a strain on our budget. In summer, the air conditioner used less, and we could keep it under $130. During mild weather when we used neither heater nor air conditioner, our monthly bill was under $100. That was better, but we had to do something.

We gradually decreased our use of electricity through specific steps, starting with making the house more energy-efficient. When I published *5 Acres & A Dream The Book*, we had just begun working on this. We were replacing our old single glazed windows and drafty doors with energy star rated replacements. We were adding insulation in walls where previously there was none.

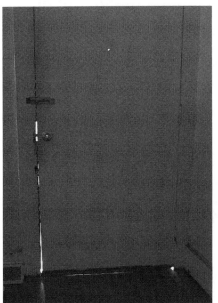 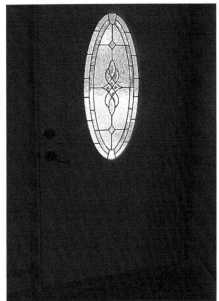

What a difference a door made. Left: original front door. Right: new front door.

We have continued to work our way room by room on this project. We bought the windows from a builders surplus warehouse and were able to purchase new windows at a fraction of the cost of buying them through a retailer. They are odd-lot windows: custom ordered extras, returns, cosmetically damaged, or missing screens. The trade-off for low prices was that we were rarely able to find replacements the same size as our originals. Nor were we able to find more than a few that matched. That meant Dan had to do some creative installation, but it also allowed him to add insulation. As a result, our house not only stays cooler in summer and warmer in winter but has a unique custom-built look which I love.

The dining room windows were original to our 1920s-built house: single glazed, cracked glass, broken latches, and uninsulated spaces for the window weights. The Energy Star windows plus insulation made a huge difference in house comfort.

Lifestyle changes have also helped us decrease our use of electricity. Much of it is common sense, such as turning off lights when we leave the room, unplugging small appliances when not in use, or putting them on a power strip to turn on as needed. Another way is using manual tools.

> *"Kitchen tools I use frequently are kept handy on a utensil rack. I realized that the tools I reach for are out of habit, so I had to retrain myself to think of hand tools first. Keeping them visible and available helped with that. Hand tools are often quicker and simpler to set up, use, and clean up than electric gadgets and gizmos."*
> *"Energy Self-Sufficiency,"* 5 Acres & A Dream The Book *(p. 115)*

We seek out hand-powered alternatives for every tool and appliance we have, and we learn how to use them. Some, like my KitchenAid mixer, are never used again. Others, like my Wondermill electric grain grinder, are still used for convenience.

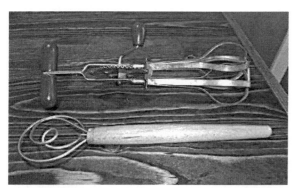

Left: An egg beater and dough whisk replaced my KitchenAid mixer.

Bottom: Country Living manual and Wondermill electric grain grinders.

Top row: Alternatives to electric coffee makers: our stovetop (or campfire) percolator on the left and a French press on the right.

Middle left: My hand-crank Vortex blender works as well as my electric model, except for liquefying garlic. Right: My carpet sweeper.

Bottom left: My wind-up kitchen clock.

It's the same for Dan with his power tools, with the advantage that manual tools don't seem to need repair as often. For equipment, we look to decrease fossil fuels as part of our energy self-sufficiency. It might take longer to do a job, but if we have the time we both appreciate the quieter, less smelly method.

Left: Gasoline sickle bar mower. Right: European scythe. Both for cutting hay. Guess which one requires more maintenance, repair, and general fiddling with.

The biggest lifestyle change has been giving up air conditioning.

> *"Because of our hot humid summers . . . I knew that if we had an air conditioner, I would use it. I would set the thermostat high, but as long as we had the electricity to power it, I would use it."*
> *"Energy Self-Sufficiency,"*
> 5 Acres & A Dream The Book *(p. 110)*

At one time I didn't think I could give up our AC. What changed my mind? Several things. The first was the cost to run it. To keep our monthly electric bill within our budget I had to set the thermostat a little above 80°F (26.6°C). The problem was that this wasn't enough to make the house feel comfortable. When the unit was running I wanted to grab a sweater, but as soon as it turned off I was hot again. Even worse, it kept the temperature difference between inside and outside wide enough so that when I went outdoors the heat seemed stifling! Eventually, we decided to not use the AC but to take other measures to cool the house.

Those other measures are pretty much common sense. During hot weather, I open the windows in the evening after the outside temperature cools down. We use box fans to vent hot air and pull in cooler air. When the thermometer climbs in the morning, I close the windows and curtains on the sunny side of the house. In winter, I keep curtains and blinds closed unless the window is receiving the sun's warmth. Then I open them.

Ceiling fans are important and account for our higher use of electricity in summer. Ours are reversible, so they gently push heat down in winter, but keep a comfortable airflow going in summer. Keeping air circulating is important because it helps deter mildew, which is a problem in the humid southeast.

Another helpful addition to the house was a solar attic vent fan. During summer, it runs when the sun is up and helps maintain a ten to fifteen degree difference between indoor and outdoor temperatures. Plus, being solar powered, it doesn't add to the electric bill.

New windows and insulation helped, but the attic vent fan made an amazing difference in the indoor temperature. Mid-afternoon in the middle of July it's 100°F (38°C) outside, but only 82°F (27.7°C) inside. A welcome difference. The fan kit came with a thermostat, so the attic retains warmth in cold winter weather.

Cooking is another area in which we've decreased our use of electricity. I use my wood cookstove in winter and my solar oven on sunny summer days for bread, rice, eggs, potatoes, tea, etc. We use the grill for meat.

Above: The wood cookstove has the additional advantage of heating the kitchen. Below: Using my solar oven has the advantage of keeping heat out of the kitchen.

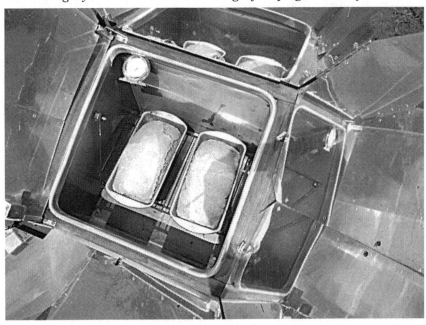

Top left: My back porch kitchen keeps cooking and canning heat out of the house in summer. Top right: A solar attic fan became our vent fan. It's connected to a sealed battery so we can use when the sun isn't available. Bottom: Hot smoking a meatloaf on the grill. Dan no longer buys charcoal, but instead uses a hickory fire.

When I use electricity for cooking, I use 110-volt appliances when I can (toaster oven, crock-pot, electric skillet, bread machine) rather than my 220-volt electric stove. Above, stew in the crock-pot and reheating rice in the steamer on top.

You might have noticed the electric dryer in the fan photo on the previous page. Yes, I still use it on occasion, but mostly I use my solar and wind powered clothesline.

In cold weather, we use our soapstone wood heater and wood cookstove for heat instead of the electric HVAC. Although we use ceiling fans to help circulate heat throughout the house, we also have two Ecofans. These stovetop fans work because of the temperature difference between the base of the fan and the cooling fins on top. Once the stove heats up, the blades begin turning and circulate heat without a plug or a battery. We've found them to be very effective.

Thermoelectric Ecofan.

In winter, we close off parts of the house we don't live in, such as the spare bedroom, the pantry, and the utility room. When we have to work in cooler parts of the house we wear more clothes.

For a number of years, I thought those lifestyle measures were probably as far as we could go toward energy self-reliance. Then I was asked to review *Prepper's Total Grid Failure Handbook* for Ulysses Press. Written by Alan Fiebig and Arlene Fiebig, it is completely unlike the dummies and idiots type beginner books I checked out from the library. Those books give a cursory nod to off-grid power but focus on somewhat overwhelming explanations of grid-tied systems. The Fiebigs' book discusses practical small-scale off-grid solar electricity.

That book reawakened my interest in going solar and I started exploring options. We were already using a solar oven, solar attic fan, solar shed lights in the barn and workshop, and a solar energizer for the electric fence. Was it now feasible to actually get off the grid? When I wrote *5 Acres & A Dream The Book*, an online calculator estimated the cost for us as being over $71,000.[1] Updated calculations estimated a cost of roughly $30,000 plus taxes, freight, and installation. Unfortunately, even at that tremendous drop in price, the idea is still out of financial reach for us.

What about a grid-tied system? Aren't there still tax incentives for that? Couldn't we offset the cost by selling our surplus electricity to the utility

Solar shed light in the old milking room.

company? This option is always very appealing, but after careful research, my answers to those questions are no, yes, and maybe.

The two main draws to a grid-tied system are tax rebates and selling surplus electricity. Unfortunately, tax rebates don't benefit people like Dan and me. To receive a rebate we'd have to buy the system first, and then apply the rebate when we filed our taxes. This is different from a grant, which is applied directly to upfront costs for the system. Even so, you might well be wondering if selling our extra electricity back to the utility company would help cover the cost.

I went so far as to obtain a copy of the contract we'd need to sign with our utility company to go grid-tied. That was a real eye-opener, and I'm going to state that if you are considering doing this, the contract needs to be read *very* (repeat, *very*) carefully beforehand. Our utility company limits the size of the system to 90% of one's current usage, requires considerable extra homeowner's insurance to protect the grid, and gives itself permission to inspect any part of the homeowner's system (indoors and out) at any time with "reasonable" notice. The contract also includes a few interesting clauses that are worth noting. One is that they agree to purchase the customer's excess electricity but are not required to do so. Another is that they can require the customer to interrupt or reduce production. And, of course, if something goes wrong it isn't their fault, and they are not liable.

Many utility companies now add monthly solar usage fees to offset their profit loss. All the above add up to a big no-way to the question of possibly tying into the grid.

Prepper's Total Grid Failure Handbook, however, gave me the idea for the plan I mentioned in chapter six. If putting the whole house on solar wasn't feasible, couldn't we use a smaller sun-powered system for our refrigerators and freezer only? This follows our "food first" motto in prioritizing projects, and it would be a relief to keep these appliances running during a power outage. Admittedly, this doesn't occur often, but it does happen. Summer concerns are hurricanes and lightning storms; in winter it's ice storms. These are real possibilities in our part of the country and could mean extensive food loss with a prolonged power outage.

To get a grasp on some solar basics, I started with a small project—a DC (direct current) powered box fan. My little system included a 12-volt 35 amp-hour sealed deep cycle battery, an 18-watt solar panel, and a simple charge controller. It proved to be very handy and I learned useful things, like how to monitor the battery. But I also learned that an 18-watt panel wasn't enough to charge the battery quickly and that 35 amp-hours doesn't last very long. However, this baby step project was a start, and while jumping into a larger project was still daunting, it gave me a basic familiarity with the components and how they work.

With that under my belt, I started to think about our larger solar project. My first questions revolved around the number and size of solar panels and batteries. I found the solar panels on Craigslist, leftovers from a larger job. We were able to get several new 345-watt panels for $240 each.

My first DIY solar experiment was a small 3-speed 12-volt room fan. In addition to the fan, the system required a charger for a deep cycle battery and a 12-volt adapter plug socket. The adapter is available with either clamps or eyelet terminals to connect to the fan to the battery.

What I needed to know was how many panels I'd need and how to size a battery bank to run my fridge and freezer for several days of no sun.

My first step was to measure how much electricity these appliances use. To find that information I used a Kill-A-Watt meter (pictured right). I measured each appliance separately and added them together. Here's what I learned.

The fridge uses 2.6 kWh/day
The freezer uses 1.6 kWh/day
Total for both is 4.2 kWh/day

Next, I looked for an online calculator into which to plug those numbers. Many solar businesses have these on their websites, usually geared

Kill-A-Watt meter

toward the brands and services they sell. They may give results as packages they offer or want you to contact them for the results. The most helpful information I found was step-by-step instructions at *Preparedness Advice* (https://preparednessadvice.com/solar/many-solar-panels-batteries-power-grid-system/). I followed their instructions, tried a couple of online calculators, and compared results. Those results were similar but dismaying.

Assuming I receive at least five hours of good sunlight daily (which is iffy), here's what I'd need to store enough energy for three cloudy days:

    1.32-kilowatt system
    4, 345-watt solar panels
    11, 200 amp-hour 12-volt deep cycle batteries
    Or 8, 260 AH 12-volt deep cycle batteries

That was more than I expected, especially since batteries cost hundreds of dollars each. Plus, I'd still need other components for the system.

I'm going to pause here to explain about the batteries. Batteries used for solar battery banks are deep cycle batteries, not cranking (starting) batteries. For a vehicle, the battery has to supply enough power to rotate the crankshaft and get an engine started. Cranking batteries must be able to bear a high draw of electricity (load) for a short duration. Solar-powered applications don't take as much energy for start-up, however, they draw continually (or at least intermittently) to keep running. A deep cycle battery bears a lighter load for a long time and can be repeatedly discharged and charged.

Deep cycle batteries are commonly measured in amp-hours (AH). This doesn't refer to actual time, because performance varies with conditions such as temperature. Rather, AH are a way to compare the relative storage capacity of various deep cycle batteries. The greater the amp hours, the longer they last before needing recharging.

Back to my results. I was dismayed because I would need more batteries than I first thought. Considering that 12-volt 200 AH deep cycle batteries start at $350 for the cheaper ones (and 260 AH start at about $500), I had a budget problem. We had $1500 for this project, which had to cover batteries, a charge controller to regulate charging the battery bank, and an inverter to plug my alternating current (AC) appliances into the direct current (DC) system. Plus, we'd need all the miscellaneous items such as connectors, fuses, racks, wiring, etc.

The pantry refrigerator, however, is an old one. Out of curiosity, I ran the numbers again, this time, with a new Energy Star fridge. I used an average estimate for low-end energy-efficient refrigerators and plugged in the kilowatt-hours from the advertised energy rating. Here's how the numbers changed.

> Energy-efficient fridge 0.94 kWh/day (compared to 2.6!)
> Same freezer, 1.6 kWh/day
> Total for both appliances is roughly 2.6 kWh/day

That's the same amount my old fridge alone used in one day! For three day's electricity storage we'd need:

> 0.8-kilowatt system
> 3, 345-watt solar panels
> 7, 200 AH 12-volt batteries
> Or 5, 260 AH 12-volt batteries.

As you can see, energy-efficient appliances make a huge difference! But it wasn't enough to keep the cost of the project within our starting budget. For the time being, feasibility remained near impossible.

Now what? The goal was a worthy one—emergency back-up for food storage—and the idea was a good one. I needed to explore alternatives.

One of those alternatives was something mentioned in the Fiebigs' book. For refrigeration, they converted a chest freezer to a refrigerator. The advantage is that a chest freezer retains cold better than an upright appliance. Open the door of a fridge or upright freezer, and you can feel cold air pouring down onto your feet. Keeping that cold air in the fridge would mean it would run less.

The freezer-to-fridge conversion option was questionable for two reasons. One was the additional cost of the freezer which further pushed our budget restraints. The second was that I couldn't find information on the electricity usage of the unit as a fridge to plug into my calculations. I could find it for the freezer, but surely it would take less to power it as a fridge.

The whole project seemed to have come to a dead-end. What came to mind was something I've told my readers often over the years—something is better than nothing.

> *"It is better to grow one potted tomato plant on the patio than none at all. It is better to have a small suburban garden than none at all. It is better to keep a few potted herbs under a grow light than none at all. It is better to do something rather than nothing."*
> "Food Self-Sufficiency: Feeding Ourselves,"
> 5 Acres & A Dream The Book *(p. 81)*

This time when I crunched the numbers, I did it for the freezer only. Then I asked myself, if we can't afford enough batteries for three days, how many can we afford? I reckoned that if I got only one day of extra freezer running time, that would be more than the grid would supply if we lost power.

We already had the panels, purchased off Craigslist from a contractor who was selling leftovers from a recent job. They were new but also heavily discounted; that helped. The other main components would be a battery bank, a charge controller to monitor and regulate the charging of the batteries, and an inverter to change the DC (direct current) electricity stored in the batteries into AC (alternating current) electricity for our freezer. We used our $1500 savings to buy these three items. We knew there would be additional costs for wiring, connectors, circuit breakers, etc., but we could purchase them out of our household budget as we needed them to build the system. "Pay-as-you-go" enabled us to cover the rest of the cost.

*Above: We bought the solar panels first, so they went up first.
Below: Dan made an adjustable rack with materials he had on hand.*

ENERGY SELF-SUFFICIENCY 117

Top right: The panel array rack is made from home-milled lumber. The top of the solar panel frame is attached to it with door hinges.

Bottom left: The bottom of the frame is attached with angle iron, metal pins, and strut channels.
Bottom right: Metal pins in the rack's lower horizontal brace enable the panels to be angled by changing the pin's slot in the strut channel.

Left: So how do we adjust the angle of the panel array? We have a nail glued to the small panel on our portable solar charging station (see the first photo at the beginning of the chapter). On a sunny day at noon, the smaller panel is tilted so that the nail casts no shadow (circled in the photo). Then the array angle is adjusted to match that angle by changing the slot on the strut channel. The reason we didn't put the nail on the array itself, is because it would stick out about thigh height. We didn't want it scratching anyone passing by.

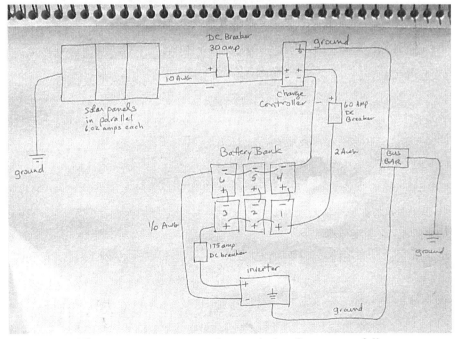

The next step was wiring the panels, but first, we carefully diagrammed the system and labeled each wire and component.

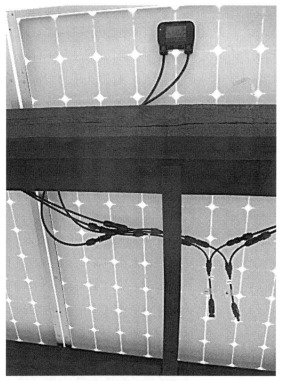

Left: We had two options for wiring the panels together: in series (negative to positive and positive to negative) or parallel (negative to negative and positive to positive). Series wiring multiplies voltage and can use smaller gauge cables. Parallel multiplies amperage but requires thicker (more expensive) cables. But it has the advantage that it will produce electricity even when the array is partially shaded. We bought an adapter kit to wire the panels in parallel.

Below: Our cable was rated for direct burial, but Dan opted for a conduit because the trench went under the driveway to the back porch.

Above: The location for the battery box is close to the freezer (located on the back porch above the crawlspace door). It is outside, which is important for battery off-gassing, and it's accessible for battery inspection and maintenance. It's shaded all day, which will help keep the batteries cooler in hot weather. Because it's only 30 feet from the panel array, we didn't need a large cable size.

Left: In summer, Dan puts a vent fan in the crawlspace door, which means we can lift the box lid and blow crawlspace-cooled air across the batteries. That's a plus considering how hot our summers can be.

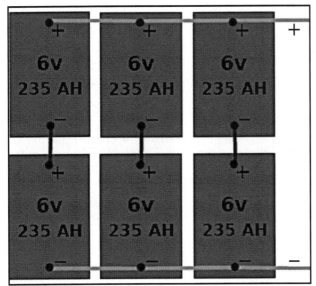

Left: As with the solar panels, we had options for wiring the batteries together: in parallel or in series. We used a combination. Pairs of 6-volt, 235-AH batteries were connected in series to double the voltage. These three strings were then wired in parallel to add amperage. This gave us a 12-volt, 705-AH battery bank.

Above: Inside the battery box. Everything is wired as on our diagram on page 119.

Next page: *We needed three circuit breakers, One for the solar panel array, one for the charge controller, and one for the inverter. These were sized according to owner's manual specifications, as were all the cable sizes.*

Bottom right: *The largest cables were needed to connect the batteries to the inverter (exiting through the conduit on the left). This photo also shows the bus bar which is used for earth grounding the charge controller and inverter.*

Grounding. Typically, elecricity follows the wiring between the source and whatever requires electricity to operate. If there's a problem, it can jump the wiring and electrify nearby metal. This could be deadly to someone accidentally touching it. Physically connecting the wiring to the earth (earth grounding) gives the electricity an alternate pathway, if needed.

Many setups ground the system as a unit. Our panel array is grounded separately, as specified in the installation manual.

Above left: Lay-in lugs connect the neutral grounding wire to the solar panels. Right: Because we grounded the panels separately, two ground rods were necessary. Below: A DC cooling fan helps vent the battery box as needed.

Left: Remote Temperature Sensor monitors the temp inside the box and enables the charge controller to optimize battery charging according to temperature.

How much did the project end up costing us? I've broken it down into our spending groups. All totals include shipping and taxes.

Major components

Solar panels, three 345-watt Sunpower photovoltaic (PV) panels	$720
Charge controller, Outback FLEXmax 60-amp, MPPT	$431
Inverter, Power Tech-On 1000-watt pure sine wave	$150
Golf cart batteries, 6 Rolls Surrette, 6-volt, 235-amp-hour	$880
	Total $1461

We chose lead-acid batteries because they are the least expensive. The trade-off is that these have the shortest life span, typically five to seven years if well taken care of. Sealed batteries can be double in price, but with the same life expectancy. Lithium batteries can double the life span, but the price is more than doubled; thousands of dollars per battery instead of hundreds. Nickle-iron batteries have the longest life, 30 to 50 years, but each battery is likely to cost in the ten-thousands of dollars.

Why did we go with 6-volt batteries instead of 12-volt? One reason is that I was unable to source 12-volt deep cycle batteries locally and shipping for batteries is very high. Marine and RV batteries sold around here are dual-purpose, listing cranking amps and low amp-hours (usually 35 to 55). They aren't cheaper and I'd need more to get more amp-hours. Plus cranking batteries won't take as many recharges as true deep cycle batteries.

Miscellaneous components

Panel extension cables, pair (40-ft, 10 AWG) with MC4 connectors	$53
Kit for parallel wiring of panels	$28
Grounding rods, clamps, and wires	$50
Wiring cables, lugs, bus bar	$136
3 Circuit breakers	$95
Remote temperature sensor	$42

DC vent fan for battery box	$12
MATE2 remote system display and manager for charge controller	$227
	Total $643

Home-built from available materials

Rack for solar panels	free
Battery box	free

Grand Total $2824

There are several ways to analyze cost. A common way is to ask how many years it will take for something to pay for itself. A low-end off-grid system might cost roughly $35,000 not including shipping, installation, and interest if buying on credit. Neither does that include a generator for backup. According to the U.S. Energy Information Administration, the average American home electric bill was $111.67 per month in 2017.[2] Do the math, and you'll likely agree that it takes more than a monetary advantage to go solar. For most people, it's a sense of environmental responsibility.

For Dan and me, the motive is food preservation. Since we rely more on what we grow than on a grocery store, this is important. If I calculate the system paying for itself in terms of food budget savings, then it will pay for itself in less than six months. Savings on the electric bill will be a bonus, as will keeping these appliances running during a power outage.

I need to add here that the cost of building the system is only one financial consideration. Eventually, parts have to be replaced, so the replacement cost should be taken into account as well. I'll use the battery bank as an example since it is usually the first thing requiring replacement.

Our solar panels should last 25 to 30 years, but flooded lead-acid batteries average about five years; longer if we take good care of them —shorter if we make mistakes. The batteries will probably be the first thing we need to replace, so we must have the money available when that time comes. Our income is low enough that we must budget for everything, so I need to take this into account now.

What am I looking at for replacement cost? Our six batteries totaled $880, with $100 of that for the core charge since we didn't have old batteries to trade in. If the batteries last 5 years, and I want to have $780 available for replacements, then I need to save $13 per month. Because prices always go up instead of down, it would probably be wise to bump that up to $15. If we want to upgrade the battery bank—in terms of battery type, amp-hours, or both—then we need to set aside more.

Other parts need replacing eventually too, although not as frequently as batteries. In addition to solar panels, charge controllers, inverters, and generators to feed the batteries during long sunless periods, all eventually need replacing. The bottom line is that even when the system is paid for, solar energy still isn't truly free.

After all the excitement, anticipation, and build-up, solar power day was exceedingly uneventful. Switching our deep freezer from the grid to our solar electric system was just a matter of closing the circuit breakers, moving the plug from one socket to another, and turning on the inverter.

Then it was on to the next step: replacing the old energy-guzzling upright fridge with a low-energy chest fridge. To do this, we purchased a 5-cubic-foot chest freezer and installed a refrigerator/freezer thermostat.

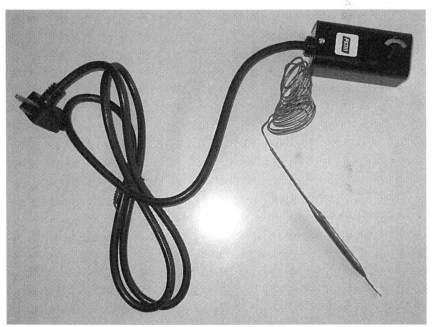

This thermostat enables me to control the temperature in the chest freezer unit and turn it into a refrigerator.

With the help of a my Kill-A-Watt meter, I learned that the unit surged to 80 watts when I plugged it in. I used a digital probe thermometer to adjust the setting, and found that as a refrigerator, the unit uses only 0.08 kWh per day—a *huge* difference from my old pantry fridge which used 2.6 kWh/day! The chest refrigerator uses less electricity in one month than the old fridge used in one day.

Even so, there are some caveats, criticisms, and a couple of challenges I should share with you. The first is that refrigeration experts say such a

The thermostat probe is placed inside the freezer. The freezer is plugged into the thermostat and the thermostat is plugged into the inverter. It was that easy!

The chest fridge is my auxiliary fridge for surplus eggs, milk, perishable produce, grains, and grain products. The ice bottle helps conserve electricity even more.

conversion isn't a good idea because a freezer compressor isn't built to operate at refrigerator temperatures. That means appliance longevity may be compromised. Another problem is condensation inside the unit, since it is operating at temps above freezing. (In a freezer this becomes the ice that needs defrosting.) This is especially true of units which have the compressor in a box that forms a shelf, rather than in the freezer walls. The interior must be wiped out regularly.

Another criticism is that bending over to get food items is considered an inconvenience. It also requires lifting items on top to get to items on the bottom. Some people can't or don't want to bend over, others think it would be too much of a hassle finding things—although I can't say that an upright with shelves is any easier. The thing I'm looking for is always shoved to the back on a different shelf! The trade-off would have to be the desire for considerable savings in electricity and a willingness to develop new habits.

For me, the biggest challenge was how to organize it as conveniently as possible. Most people use the milk-crate style file storage boxes. It took some experimenting with various crates and boxes to work out a system. All things considered, I'm very pleased with the unit.

How has our small solar system affected our electric bill? At the time of this writing, we've gotten two electric bills with both appliances off the grid. Our average usage for those two months was 11 kWh per day. To put that into perspective, back when we used the HVAC for heating and cooling, we averaged about 25 kWh per day. After we quit the HVAC in 2015, our average usage dropped to 18 kWh per day. While 11 kWh per day is a huge improvement, our long-term average will be higher. This is because of our use of fans in summer, as I mentioned earlier, and because I use my electric stove more in summer for canning.

How long does the charge in the batteries last? The bank stores about a day and a half of electricity for the two appliances. But since the panels still produce a small amount of electricity on cloudy days, and our charge controller optimizes that, I can't simply count cloudy days. I have to keep an eye on the battery bank to make sure we don't drain it too low. I do that by monitoring the batteries' state of charge (SOC).

Technically, the state of charge refers to the specific gravity (i.e. chemical composition) of each battery cell. However, the bank voltage gives an approximate idea of what's going on. It's measured when the batteries are neither being filled nor powering something, so I usually check them at bedtime. Typically, folks say not to let the batteries drain below 50%. However, they will last longer if not drained even that low. I keep my eye out for a battery bank voltage reading of 12.4. According to our battery manufacturer manual, this is about 75% SOC. If the voltage is that low at bedtime when nothing is running, then I turn off the inverter for the night and wait for the batteries to refill the next day. A refrigerator is said to keep things cold for twenty-four hours with no electricity, and a freezer for three days. As long as I don't open them, I have a time window to work with. If it's darkly overcast the next day and the panels aren't putting out much, I plug the appliances back into the grid. They are on a power strip, so if I need to plug into the grid it's easy to do. If the grid is down, well, then I have to move on to Plan B, just like everybody else.

Lessons Learned

One lesson we've learned here is never to assume nothing more can be done. Of our original goal of energy self-sufficiency, I always assumed total success—getting off the grid—was next to impossible. At the time it didn't occur to me that we would be able to make changes and discover new ways of doing things.

The other lesson became obvious when I asked myself if energy self-sufficiency is still a goal. I had to say yes but not in the same sense as we originally thought. Now, I think of the goal as not being so dependent on modern energy that my world will fall apart if it's disrupted for any length of time. Now the goal is to take whatever comes in stride, rather than worry about a wild ride. That points less to technology and more to lifestyle. And that seems to be the direction we've been heading all along.

Notes

[1] Leigh Tate, *5 Acres & A Dream The Book* (Kikobian Books, 2013), 119.
[2] Amanda Dixon, "How Much Is the Average Electric Bill?," *SmartAsset*, July 9, 2019, accessed February 10, 2020, https://smartasset.com/personal-finance/how-much-is-the-average-electric-bill.

CHAPTER 9

WATER SELF-SUFFICIENCY

"Water conservation, both greywater recycling and rainwater collection, is a step toward water self-sufficiency as well as stewardship."
"Water Self-Sufficiency," 5 Acres & A Dream The Book *(p. 131)*

Our house is connected to municipal tap water, and that doesn't lend itself to water self-sufficiency. Digging a well may never be feasible, so for now, we focus on water conservation. We've done that through rainwater collection and a conscious effort to not be wasteful. In addition, we have experimented some with greywater irrigation. In chapter six, I told you how I'm using hügelkultur swale beds in the garden to capture rainwater. In chapter ten, I'll share how we're building soil for increased water retention.

Our original rainwater collection system is the one featured in *5 Acres & A Dream The Book*. We stacked two 275-gallon water totes to collect roof run-off from the house, with two more for overflow. While 1100 gallons may seem like a lot of water, I found that during a long summer dry spell it wasn't enough. I found myself having to ration that water before it rained again.

We priced larger collection tanks, but most of them ran between $1 and $2 per gallon. Considering that the totes ranged between 15- and 25-cents per gallon, the higher prices for larger tanks didn't sit well. Diligence pays off, however, and we eventually found a 1550-gallon water tank for $800 at Tractor Supply Company.

1550-gallon water tank. To move it into place, we rolled it on its side.

Dan set it up to catch roof runoff from the house and moved three of the 275-gallon totes to the top of the garden to hold the overflow. That gives me over 2600 gallons of water for the garden.

When we built the goat barn, we discussed collecting its roof runoff for the livestock. The appeal for this was compelling; no more toting buckets of water from the house for the critters! We purchased a second 1550-gallon tank for the barn roof and set up our two remaining 275-gallon totes to catch runoff from the milking room. Because the water was going to our animals, Dan wanted to make sure it was properly filtered.

Rainwater collection from the barn and milking room yields 2100 gallons.

Our first catchment system didn't have a filter. It had a clean-out plug to collect debris from initial roof runoff, but we found that debris collected at the bottom of the tanks over time. Plus, the plug was difficult to get to and messy to clean out.

When Dan set up our large 1550-gallon tank, he experimented with a tank-top filter. He cut a 55-gallon drum and filled it with layers gravel: small on the bottom and large on the top. This worked well; no more debris in the bottom of the tank. But the filter was large and very heavy. It required a stand to hold its weight. That made it hard to clean out too.

No filter, just a clean-out plug (top right).

Dan's first experiment with rainwater filtration.

For the barn tanks, Dan tried an inline filter. An inline filter is placed inside the horizontal PVC pipe carrying water from the gutter (below). To make it, he used tubes of filter cloth, which is used to cover perforated

drainage pipe to keep out soil and debris. He filled the tube with two sizes of gravel and placed it in the 6-inch PVC pipe carrying rainwater to the tanks. Unfortunately, the filters slowed the flow of water into the tanks, and Dan wasn't satisfied with the set-up. That led to more experimenting.

The system had to be inexpensive to make and easy to maintain. Modern rainwater filtering systems in the U.S. are complicated and costly, with very few DIY options to be found. Instead, Dan turned to books in our homestead library: *Five Acres and Independence*, *Handy Farm Devices and How to Make Them*, and *Homemade Contrivances and How to Make Them*. First published in the early 1900s, these books contain practical project ideas for farmers to construct themselves.

For his new rainwater filter, Dan simply connected the barn downspout to a small gravel-filled bucket on top of the tank (upper right). This did very well to kept leaf debris out of the tank and is easy to clean out.

He also experimented with a secondary water filter (bottom right). Tank overflow drips into a 5-gallon bucket filled with sand. A hose bib at the bottom of the bucket makes it easy to fill a water pail.

Then we developed another problem—mosquito larvae in the rain tanks! Just what we wanted, a mosquito breeding farm! We were clueless as to how they were getting there, until one day when Dan checked the inside of the 1550-gallon tank. He took off the access cap and discovered a nearly drowned chipmunk paddling around inside the tank! He offered it a board to climb on, and the thoroughly drenched 'munk hightailed it out of there. It was extremely providential that he found it when he did because a dead animal in the tank would have been a horrible state of affairs. But how did it get into the tank?

We figured it had to be through the gutter-to-tank PVC pipe. With that, we realized this was probably how mosquitoes were gaining access to the water in the tank as well. Dan took apart the pipe, covered one end with fiberglass window screening, and rejoined the pieces. Problem solved.

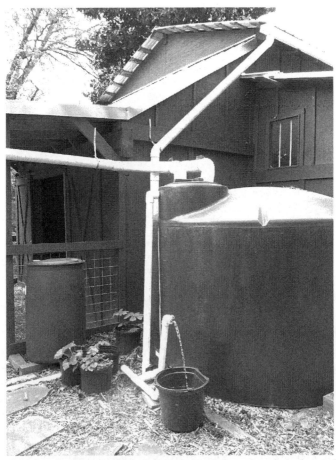

The barn tank receives water from the hayloft roof and the barn door overhang. The chipmunk must have entered the PVC pipe at the gutter.

Another problem we've had with our rainwater tanks is algae. The tanks are translucent and even though we chose shady places for them, it only takes a small amount of sun to stimulate algae growth. For the garden, this was only a minor concern. For the animals, it was a problem we needed to correct.

Opaque tanks are available, but replacing all of our tanks was not a practical option. The alternatives were either covering the tanks or painting them. The paint would have to be suitable for plastic because any other kind of paint would peel. That's what we ended up doing. We emptied the tanks, scrubbed them with soap and bleach, and painted them. So far, the paint seems to have done the trick.

The rainwater collected from the milking room roof goes through several uses. Dan uses it to water the chickens and goats, and to fill a small duck pool. When the pool water needs changing, he drains it to water our compost and the chickens' grazing beds. To keep mosquitoes out of the duck pond, we use a product called "Mosquito Dunks." This is an inexpensive, once-a-month, non-toxic product that contains *Bacillus thuringiensis israelensis* (BTI). BTI is a bacterium that kills mosquito larvae but is harmless to both animals and humans. It's approved for organic production.

We gradually continue to expand our rainwater collection and usage. Watering the garden and livestock are the priority, but using it for flushing toilets and doing laundry are other possibilities.

Laundry points to something else we've experimented with a bit —greywater. Our first idea was to use the washing machine water in a filtration bed alongside the house.

> *"For laundry greywater, I'd like to build a raised soil filtration bed alongside the house. Pergolas would frame the bedroom windows and in the bed, I'd plant deciduous vines to grow up pergolas. This will shade the windows from the afternoon summer sun, yet allow sunshine to fall upon the windows during winter."*
> *"Water Self-Sufficiency,"* 5 Acres & A Dream The Book *(p. 136)*

The holdup on this project has been replacing the windows in the front bedroom. These are some of the last windows needing upgrading, but when it comes to prioritizing projects, this one never makes it to the top of the list.

When the summer of 2016 turned out to be the hottest and driest we experienced since we started homesteading in 2009, our thoughts turned again to greywater. With weeks of daily highs topping 100°F (38°C), it was a real challenge to keep the garden alive. This was before we installed our large collection tank, so by mid-July my smaller water totes were empty and there was no relief from the heat and drought in sight. I couldn't help but think of all the tap water we were wasting down the drain. Greywater on the garden isn't usually recommended, however, so I had to do careful research on what we could and couldn't do.

There are two potential problems using greywater. One is that it quickly becomes blackwater. If allowed to sit it becomes a breeding ground for putrefying bacteria. This is why greywater can't be stored. A further safety precaution is to not use it for root crops or low-growing greens that are eaten fresh. Greywater is frequently used to irrigate fruit trees, so I used it for tall plants like corn and okra. When we don't need it, a valve allows us to redirect it to the septic tank.

The other problem with greywater is the detergents and cleaners it contains.

> *"Although some chemicals in these products may be biodegradable and considered safe for humans, they can still be toxic to plants. . . Another problem is that most soaps and detergents raise the greywater's pH, making it more alkaline. If the greywater is to be used for irrigation, this will be a problem for acid-loving plants."*
> "Water Self-Sufficiency," 5 Acres & A Dream The Book *(p. 136)*

Part of my research was for greywater safe products I could buy locally. Here's what I found.

Laundry. My first greywater safe find was Ecos Liquid Laundry Detergent. It cost me about ten cents per load, and I like that it's liquid; no undissolved powder in my laundry water. Later, I found soap nuts and like them even better. They're biodegradable and cost me less than two cents per load. Plus, the only packaging is a cardboard box. Why didn't I go with homemade detergent? Because it's high in sodium and boron, and not cheap to make. It cost me about forty-five cents per load, and the powder doesn't dissolve in our cold water.

Bleach. I rarely use bleach, but I learned that hydrogen peroxide-based bleaches (oxy types) are considered greywater safe. The homemade version is easy and inexpensive. It can be poured or sprayed onto stains before laundering.

When Dan repaired our old carport, he extended the roof. We use the area to store firewood and for an outdoor laundry area. The tank collects run-off from the carport roof. A hose drains the laundry tubs to wherever we want the greywater to go. I still use my electric washing machine, but we have this for backup.

    2 parts water
    1 part hydrogen peroxide
    1 part baking soda

Dishwashing. Most greywater experts agree that liquid castile soap is greywater safe, with Dr. Bronner's as the most recommended (probably because it's the most well-known). That one is hard to find locally, however. Dawn classic blue dish detergent has been used to wash waterfowl rescued from oil spills, but it doesn't claim to be greywater friendly. It has the fewest additives, so for now, it's the one I use.

Shampoo. It was interesting to learn that shampoo has only been around since the 1930s. Also, that at their most basic ingredient level, shampoo and dishwashing liquid are the same. Many shampoos claim to be "natural," even though they contain quite a few artificial chemicals. "No Poo" methods work very well for me. Many people use baking soda to replace shampoo and apple cider vinegar as a conditioner. Because my hair is so dry, most of the time, water-only works well. After I stopped using shampoo, I discovered that I no longer needed to buy conditioner. That meant two less greywater unfriendly products going down the drain.

Household cleaners. I use vinegar, dish soap, elbow grease, hydrogen peroxide, and a commercial scrubbing powder called Bon Ami Powder Cleanser. The Bon Ami replaced a DIY scrubbing powder that I've really liked, but it contained equal parts of table salt (sodium chloride), washing soda (sodium carbonate), and borax (boron). My first soil tests came back high in sodium, however, so I didn't want to add salt to my greywater as well. Bon Ami contains smaller amounts of these: powdered limestone (calcium carbonate), feldspar, soda ash (sodium carbonate), baking soda (sodium bicarbonate), and a biodegradable surfactant.

Body soap. So many of these contain fancy additives. I stick to those with the least ingredients. Aged homemade soaps are the best choice. The salt added to harden the bars is a small amount compared to all the sodium-based chemicals found in commercial products.

Those are the major changes I made; not perfect, but a step in the right direction. Really, anything we put on our bodies goes down the drain as well: toothpaste, deodorant, hair styling products, skincare products, mouthwash, body lotions, etc. Even hand sanitizers eventually get washed off. The same is true of other cleaning products that may get rinsed down the drain when we wash out mops and cleaning rags. I'm continually on the hunt for products that help us be better stewards of our water.

On occasion, I use small amounts of products that are neither soil nor plant friendly, such as diluted chlorine bleach as a disinfectant for poopy goat water buckets, and to periodically sterilize milking equipment and udder washing rags. These are never discarded down the drain, but further diluted and poured onto our gravel driveway, where we don't want anything growing anyway.

In addition to everything I've shared so far, we continue to do little things that help conserve water. Things such collecting cold tap water in a bucket while waiting for it to heat up enough to do dishes. That usually goes into the washing machine or to water potted plants outside the back door. Cooled canning and cooking water also are used for potted plants, as is dirty livestock water.

Dan and I continue to experiment with new ideas for managing and conserving water. I'll close this chapter with photos showing new ways we're recycling and conserving water.

When Dan built the duck house for our Muscovy ducks it included a house with doors in the back for gathering eggs and a built-in pond made from a stock tank. He put a hose bib into the tank's drain plug and attached a hose for occasional draining. The dirty drainage water is used on the compost bins and grazing beds.

New this year are ollas, which are terra cotta pots used as water reservoirs. The pot's drainage hole is plugged, then it's buried and lidded. It's good for dry areas because it allows a slow seep of moisture through the porous pot walls.

Our newest water conservation addition is an African Keyhole Garden. This brilliant concept combines growing, composting, and watering into one manageable system. Roughly 6-feet across, a compost container sits in the middle. These have proven highly effective in arid areas where gardening can be a challenge. Below, ours is ready to plant. On the cover, you can see it with thriving garden plants.

CHAPTER 10

RESOURCE SELF-SUFFICIENCY

"We ought to leave things better than how we found them. We ought to make a positive impact on our environment, our communities, and one another. We ought to be moderate in consuming and with our resources. We ought to be considerate of the needs of others, and not take or use more than we need."
"Doing What We Ought," 5 Acres & A Dream The Book *(p. 196)*

I chose the title of this chapter to fit the self-sufficiency pattern of my other chapter titles. It would be more accurate, however, to call it "Resource Stewardship." Without stewardship, there would be nothing with which to be self-sufficient. There would be nothing on which to be self-reliant. That being said, I must also acknowledge that "resources" is a pretty broad term with numerous applications. Environmentalists speak of renewable and nonrenewable resources. Economists divide resources into capital, natural, and human categories. Project management looks at labor, equipment, and materials as resources. What about homesteading? Do any of these models fit my requirements?

Well, what exactly is a resource? It could be defined as:
1. A readily available source of supply, support, or aid.
2. A source or supply from which benefit is produced.
3. That to which one resorts or depends upon for supply or support.

As a homesteader, I view resources as the things that are available to help us work toward self-reliance. I acknowledge that I must manage and maintain these responsibly, with a view toward balancing supply and demand. To do this properly requires the right tools and equipment, plus the knowledge and skills to use them. Figuring out the most appropriate tools and equipment for my needs takes time and experience. So does acquiring them. Learning to use them takes practice.

Our tractor is a good example of necessary equipment, although it took us several years to get one. In our early years here we tilled, and for that, we used our tiller. That was okay for the garden, but for larger areas, such as pasture, grain, and hay fields, the tiller was too small. Farm tractors aren't cheap, so the solution we hit upon was a two-wheel, or walk-behind, tractor, pictured below. Our problem was that these are no longer manufactured in the U.S. While still popular in Europe, availability for us was as either an extremely expensive import or an older used model.

Our first farm tractor, a 1967 Simplicity Model W Walking Tractor, once sold by Montgomery Ward.

The used model we found cost us $1000 and came with several attachments. The problem was that because it is no longer manufactured, it was impossible to find parts and more attachments. So we sold it. Eventually, we found the right tractor at the right price.

Our 1960 Ford Powermaster 861. Dan haggled the price down to $3000 and got free delivery thrown in to boot.

We used the tractor to plow and disc for only a couple of seasons. When we switched to no-till, we sold these attachments, but the tractor continues to be a true workhorse for us. We use the scrape blade and boom pole, and it's enabled Dan to haul logs to the sawmill, and lumber and timber to his project sites. The power take-off (PTO) operates our wood chipper and auger for digging fence post holes.

In addition to the tractor and its attachments, other invaluable equipment we've added along the way includes our sawmill (pictured page 26), welding machine, pole saw, and sickle mower and scythe (both shown on page 106). There are other things that we'd like, such as a no-till grain drill, but as with everything else, we'll just have to wait and see if that can someday become a reality.

Of our resources, we initially focused on food, livestock, energy, and water. We have spent a lot of time (years!) learning how to be good stewards of these things. We're still learning. How do we view our resources ten years later? I've discussed food, energy, and water in their own chapters. Here, I'd like to focus on three more resources we've realized are foundational to our way of life: our soil, our trees, and our time.

Our Soil

> *"The number of animals we can keep begins with the soil. Healthy soil means healthy plants, which means healthy animals. It is a balance from which we all benefit."*
> *"Goat Tales: It Was Time for Elvis to Go,"* Critter Tales *(p. 138)*

If I had to choose the most important resource on our homestead, it would be our soil. It's the foundation of everything we need. The first thing that comes to mind is the garden, but also we have livestock that need good pastures and hay. In addition, we want to grow field crops such as wheat and corn, plus extra greens and root crops for the critters. The challenge is that we live in an area with regionally poor soils and uncooperative weather, especially long hot, dry spells most summers. For the garden, I can make compost and use mulch to build soil, but the pastures and field crops have been more challenging. The question was, how can we build soil over acres, not just garden beds?

> *"After reading Neal Kinsey's* Hands-On Agronomy *we decided to test the soil in our pastures, one at a time. . . It was trying to track down those amendments that became a challenge, and it took about six months to find sources for everything we needed."*
> *"Food Self-Sufficiency: Feeding Our Animals,"*
> 5 Acres & A Dream The Book *(pp. 94-95)*

We began soil improvement with remineralization. Every year we would have one of our paddocks tested for a complete mineral profile and follow the recommendations for organic soil amendments. Our plan was to focus on a different pasture each year, which meant annual soil testing and purchasing the required amendments. We started out enthusiastically, but this approach turned out to be labor, time, and financially intensive. A detailed soil mineral analysis is more expensive than the basic testing done by the state cooperative extension. I was willing to pay for it, but finding the recommended minerals in bulk quantities became discouraging. Only lime was available locally, so everything else had to be shipped in from out of state. Shipping and handling often cost as much or more than the minerals themselves, and I was spending somewhere around $500 per year for one acre. Because our budget is always tight, the money sometimes wasn't there. Then too, we had to re-test our soil and top up with more amendments every year. Where was the self-sufficiency in that? Our plan to remineralize annually eventually fell by the wayside.

But doing nothing accomplishes nothing, and our soil showed it. One day, Dan found a video on YouTube by a North Dakota rancher named Gabe Brown. "You've got to see this," he said. We ended up watching not just one video, but several. In video after video, the message was loud and clear: feed the soil microorganisms and the rest will fall into place. The results were irrefutable. Gabe Brown farms thousands of acres with roughly twelve inches of precipitation per year. He doesn't irrigate and adds no fertilizers or soil amendments. Yet, he has rich moist soils, consistently lush growth, and excellent soil test results. His secret? Carbon.[1]

This idea was revolutionary for me. For as long as I could remember, the key element for healthy plant growth was considered to be nitrogen. My organic gardening books have whole sections devoted to nitrogen, while carbon is just a blurb about photosynthesis: plants take in carbon dioxide from the air and give off oxygen and glucose. Excess glucose is stored in the roots and explains why carrots are sweet. That's where most explanations end. Now, I was learning more. Plant roots don't just store glucose, they secrete it as liquid carbon to feed soil microorganisms, especially mycorrhizal fungi.

Mycorrhizae are truly amazing. They form symbiotic relationships with plants and exchange that liquid carbon for other nutrients the plant needs. They do this by extending the root system so that these nutrients can be harvested from other areas and transported to the plant. More amazing, the fungi network with one another to extend their nutrient harvesting to areas covering acres and miles.[2]

Mycorrhizal fungi networking on potato roots.

I also learned that soil bacteria are another key player. Most of us have heard of nitrogen-fixing bacteria. These form nodules on the roots of legumes to convert atmospheric nitrogen into a form that plants can use. Other bacteria are decomposers. They are soil builders, turning dead plants into organic matter. Both kinds of bacteria feed on carbon, either directly from the plants or from organic matter in the soil.

Soil building occurs when carbon-containing organic matter, nutrients, and soil particles—sand, clay, and silt—are bound together with glomalin. Glomalin is a sticky substance secreted by certain mycorrhizal fungi. The result is soil aggregates which store carbon and nitrogen, reduce wind and water erosion, increase water infiltration and retention, reduce soil compaction, and protect and improve nutrient cycling.[3]

Soil building, then, is a biological process that requires a symbiotic relationship between living plant roots and soil microorganisms. The microorganisms feed on carbon, and they can obtain it in several ways. Their preferred choice is fresh, from plant root secretions. Their back-up choice is carbon from dead plant roots in the ground. After that, they will feed on carbon from mulch and dead plant matter on the surface of the ground. If none of the above are available, they will consume whatever organic matter the soil has to offer.[4] It doesn't take a rocket scientist to

Nitrogen-fixing nodules in a bed of cowpeas.

figure out that soil organisms will deplete soil fertility if they have no other survival option. If we want to build our soil, we have to keep these tiniest of critters well fed.

There are other benefits from soil carbon. Combined with water, it forms carbonic acid, which extracts minerals from rocks in the soil. This is the same principle used to make mineral-rich bone broth (page 196). An acid such as vinegar is added to the stockpot to slowly dissolve the minerals from the bones. Carbonic acid is nature's way of mineralizing soil.[5]

Carbon stabilizes soil nitrogen. Nitrogen is volatile and if not utilized by plants will escape into the atmosphere. Carbon is able to sequester nitrogen in the soil, i.e., keep it stable until soil microbes need it. The magic ratio is 25 to 30 parts carbon to 1 part nitrogen. That should sound familiar to all you serious compost makers out there.[6]

So how does Gabe Brown get carbon into his thousands of acres of soil? Is it possible to compost and mulch that much ground? He does it with no-till cover cropping. A no-till seed drill combined with species-diverse cover crops have built his soil. Those cover crops are grown before, with, or after his cash crops. Plant residue is allowed to stay on the soil to first mulch and then decompose to feed the soil. By adding legumes to the seed mix, nitrogen is fixed and added to the soil as well. By not tilling, he doesn't destroy mycorrhizal fungi but allows them to continue to build and maintain their vast underground network of nutrient transportation.

I immediately realized that I had been doing something similar with my modified Fukuoka planting in our pastures (discussed in chapter 7, "Feeding Our Animals"). I was doing it to improve the soil but I hadn't made the carbon connection. As a longtime gardener and homesteader, how could I have missed it? I could have lamented my lack of knowledge, but this information was exciting because of its timeliness. Dan and I had been trying to figure out a huge puzzle for several years. We thought we were doing the right things but our soil wasn't improving. That was discouraging. What we didn't know was that we had puzzle pieces missing. Finally, those missing pieces were falling into place. That in itself was a relief, but we knew that we had a huge task ahead of us. We had to figure out how to apply this new knowledge.

We needed to come up with a plan. First, we identified three main areas in regard to land usage: gardens, pasture, and field crops. Then, we decided on specific soil building strategies for each.

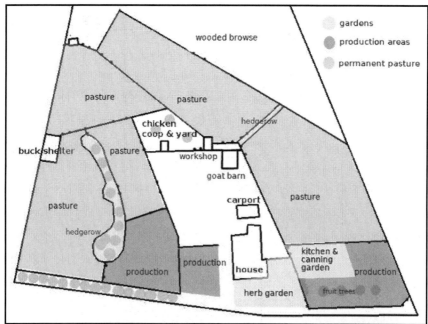

Our land usage map indicating areas for gardens, pasture, and field crop production. Field crops include grain, hay, and other feed crops.

Gardens. These include a kitchen and canning garden for annual vegetables, a slowly expanding perennial herb garden, and a couple of permaculture hedgerows. These are our smallest areas and so easiest to work with. Happily, we're already on the right track because we compost, mulch, and don't use synthetic fertilizers or weed or pest killers. The plan is to continue making huglekultur swale beds, use cover crops during times of rest, and expand companion planting for increased plant diversity.

Pastures. Our pastures aren't large in terms of acreage, but until now, the how-to of soil improvement for them escaped me. Even so, because of my modified Fukuoka planting method, this is another area in which we've done something right. In addition, we've begun a more intensive grazing rotation, which allows the forage to recover and grow. Even so, I need to work toward growing more perennial forages. In spring and fall, I hand broadcast the most diverse seed mix I can manage (See Appendix C, page 209). Then we mow the existing growth to mulch the seed. I spot seed and mulch bare areas every time I clean out the barn.

Field crops. In farming lingo, these are called "cash crops." Since we use these crops ourselves instead of trading them for cash, I prefer to call them production crops. Production areas are where we grow hay, grain, and bulk root crops. They are similar to a garden in that the crops are annuals and change seasonally. However, these plots are larger than a garden, so they need different methods for managing.

Hay field. Larger than the garden, but small enough to scythe by hand.

Hay rakes don't catch on stubble like metal garden rake tines do.

Large-scale producers such as Gabe Brown build soil on large acreage through cover crops and no-till planting. However, they have the large equipment to do so. The smallest no-till seed drills for small tractors or ATVs run about $7000 to $8000 or more, which is out of the question for us. Instead, we utilize other techniques.

1. Broadcast seed into existing forage (undersowing), then mow and leave the cuttings as a mulch (carbon) layer. This works best in hay and grain growing areas.
2. Rotate production crops with cover crops using the same undersow and mow method.

Method #1 in action. The seed was sown and then Dan cut the remaining growth with our lawn mower. Bare spots were covered with old hay (in the garden cart.)

Equipment-wise, these are the simplest methods. And even by hand, they are certainly less labor-intensive than plowing or tilling.

There are other ways to add carbon to the soil. Many folks like biochar and have recommended it to me. While I won't argue its results nor try to dissuade anyone from using it, I'm not inclined to go that route myself. It doesn't make sense to buy it if other free sources of carbon are available, nor to invest the time into making my own. For me, the simplest methods are always best.

Our Trees

Roughly half of our five acres are wooded. This was a selling point when we first saw the place, but at the time we were thinking "firewood." We didn't realize that the trees were mostly pine. Initially, that was disappointing, but as we've lived here those pines have become valuable as a homestead resource.

Pines are a pioneer species, which means they are the first trees to grow in a neglected clearing. Because of that, we wonder if our woods was logged many years ago. Our next-door neighbor recalls his grandfather plowing their property with a mule, so it's possible ours was farmed too. The previous owners of our property kept the front half cleared and allowed the back to grow. By the time we bought it, it had grown into what we call "the woodlot," with many mature pines and a number of young hardwoods.

Our first year here I created a pleasant walking trail through our woods. Then the pines started falling randomly. Some of them broke off mid-trunk and fell. Others were completely uprooted. It seemed that every time we checked, at least one more pine was down. My walking path was obliterated, and our peaceful woods became a hazard zone.

Pine trees breaking mid-trunk or uprooting to create a mess.

I spent many hours researching causes. We discovered we had a pine bark beetle infestation (common in our part of the country), but I thought perhaps the problem was disease. What else could weaken trees? Then I learned about ecological succession. As pioneer trees, pines grow and mature more quickly than hardwoods. The slower-growing hardwoods come up in their shade. With their shorter life spans, pines are at their end-of-life by the time the hardwoods require sun. Our pines were large and old. Many of them were spindly due to competition for sunlight. What we were witnessing was a natural woodland succession.

We reluctantly accepted what was happening, but several things bothered us. One was the loss of our beautiful mini-forest, the other was the waste. Trees were lying everywhere and it seemed there ought to be some use for them. Pine can't be burned in a wood stove, but many of those fallen trees were solid and could be useful as lumber. This was also true of the pines that remained standing.

Dan's first thought was to have someone come in to log them. There would be a logistics problem because there is no road access to this part of our property, but we didn't know what else to do. However, no one was interested because our tree lot was only a couple of acres. The professionals didn't log wooded areas under ten acres.

For a time we did nothing. The oldest pines continued to fall, often taking out fences. That meant escaping chickens, goats, and pigs, plus repair. The trees fell unexpectedly, but especially on windy and stormy days. We kept ourselves and the animals out of the woods on days like that.

When we began to make plans to build my goat barn, Dan would occasionally remark how useful it would be to have a small sawmill. I showed you his chainsaw mini-mill in *5 Acres & A Dream The Book*,7 but using it was slow going and rough on the saw's chains. Chainsaws typically come equipped with crosscut chains. These are used to cut down trees or cut branches and logs into sections, i.e. across the wood grain. To cut the length of a tree or log, a ripping chain is better. Ripping chains are made to cut with the grain and have teeth set at a different angle. But even with a ripping chain, the mini-mill was best suited for a few boards or posts. Sawing a whole barns-worth of lumber wasn't realistic.

On a whim, we took a look at Craigslist and found a small portable sawmill for sale. Providentially, Dan happened to be home from work that day, the seller was home from work that day, the saw was priced to sell, and we had enough money in our barn savings to get it.

Dan started milling the fallen pine trees first, many of which were hung up in other trees. He also cut down and milled trees that leaned

Milling a 6-inch by 6-inch post for the goat barn.

precariously over fences. There was a modest learning curve to using the sawmill, but with practice and experience, he improved quickly. He milled all the timbers and lumber for our goat barn The edge slabs and scraps are useful too. They've come in handy for knee braces, pegs, benches, garden bed borders, stakes, and boards for walkways.

Left: Freshly milled 2x4 boards and a 6x6 beam. Small slivers of wood (called "stickers") are put between curing lumber to allow for air flow.

Barn bench made from mill waste slabs of fallen pine trees.

Besides lumber waste, there are numerous twigs and branches from each tree; enough to make a huge pile that can't be milled and aren't good for burning. However, they are perfect for making wood chips.

Before we got the sawmill, we would rent an industrial chipper once a year to chip our numerous fallen branches and wild shrubs. As we worked we would say, "If only we had our own chipper." The one we eventually got was a small backyard model; only good for leaves and twigs. I showed you what we did with it in chapter seven. Instead, we set our sights on a larger model.

Research, patience and saving a little money each month paid off, and finally, we were able to buy a midsize PTO-driven wood chipper.

PTO-driven means it operates off of our farm tractor's "Power Take-Off." In other words, it takes its power off of the tractor's engine. It can chip branches up to three or four inches in diameter and so is useful for twiggy branches from firewood cutting plus scrap and waste from the sawmill. Those wood chips are an important resource for us.

Most of the chips become mulch, but they make a good ground cover in other ways. The goat yard for example. There, they help keep dust down in dry weather and cover the mud when we get a lot of rain. Of course, the goats add manure and urine, and eventually, the chips need to be replaced.

Chipping day starts by cleaning out the goat yard and putting the soiled wood chips in the compost bins in the chicken yard. The chips supply carbon for the compost and the manure and urine supply nitrogen.

WoodMaxx WM8M PTO-powered wood chipper.

The chips decompose with the help of kitchen scraps and our chickens. I find wood chip compost to be superior to the compost I used to make from soiled straw from the barn. Mycorrhizal fungi love decomposing wood, and I don't have to worry about grass seed sprouting in the garden.

The chipper was an expensive piece of equipment, but like the sawmill, it has more than paid for itself; not only in wood chips but in helping us clear the tree debris from our wooded areas. It meets needs on many levels and points to the importance of investing in the right equipment to manage one's resources.

As the pines have been cleared out, our young hardwoods have started growing. Because these are our source of firewood, we want to manage them properly. For every hardwood cut down, we allow one or more new hardwoods to grow in its place. Or we plant one. With careful stewardship, trees are an excellent renewable resource.

Our Time

'Things start well but soon become overwhelming: too many projects, too little time, too many things going wrong. The workload gets heavier, the to-do list gets longer, and there are never-ending demands on your time and energy. Things aren't working out the

Dirty wood chips from the goat yard go to the compost piles. We add kitchen and garden scraps, and the chickens help turn it into rich, black compost.

way you expected and the dream has become a nightmare."
"Keeping Things Manageable,"
Prepper's Livestock Handbook (p. *190*)

Time is a non-renewable resource. In terms of its whole, we have an unknown quantity. Because we live in time, we think of it in terms of units —years, months, weeks, days, hours, etc. Does it pass at a steady rate? Often it doesn't seem that way! How we view its rate of passage is usually tied to the things we think we need to do; busyness makes time go quickly, boredom seems to slow it down.

Time can be used constructively (a relative concept) or it can be wasted. Wisely managing time is no less true in homesteading than any other aspect of life. For Dan and me, that has meant learning how to work smarter, not harder.

Our late start in homesteading has been a source of time frustration. This was especially true when Dan was in the workforce and only home a few days each week. Job retirement has enabled him to be home full-time, but the habit of feeling time-pressured hasn't been easy to shake.

> *"I think if we were younger when we started, we might have prioritized differently. . . . As it was, the younger half of our lives was behind us, and we didn't feel we had the time to gradually build our homestead."*
> "Setting Priorities," 5 Acres & A Dream The Book *(p. 28)*

Part of the problem, I think, is social conditioning. We are taught to evaluate our time in terms of dollars per hour. We're told to "spend" our time wisely and not "waste" it. In that regard, society has a somewhat dualistic view of time. There are only two acceptable things we can do with our time: work or play. Anything else is considered wasting it. Taking time to smell the roses sounds good, but in reality, it just doesn't fit in. It took me a while to simply sit and enjoy watching my baby goats at play without feeling like I ought to be doing something else.

Time also seems to be used as a measure of success. For many people that might mean their ratio of work versus play, because successful people are viewed as having more time for entertainment and leisure. For Dan and me, that success might be evaluated as completing a project "on time," and delays are seen as obstacles. But when our minds are geared this way, delays become frustrations. Yet who can control the weather, for example, or the other factors which influence the completion of a project?

I've been asking myself, lately, if it's possible to view time differently. As a tool, it's useful for making appointments and planning shopping trips when the stores are open. On the other hand, our critters don't care if we just switched to daylight savings time. Their biorhythms determine when it's time to eat, not a clock. Twice a year when the time changes, I find myself having to stop and calculate the "real" time for my morning chores.

What I am beginning to realize is that time is a gift. Each day is given to everyone in equal measure. We each have twenty-four hours in a day to do with as we choose. The question is, am I choosing wisely?

How I answer this question points to how I view my sense of purpose. The age-old question of "why am I here?" is no less important than the

Working smarter not harder means the right equipment. Dan made this small wagon from a broken riding lawn mower frame and steel angle iron from the wood chipper's shipping crate. The only purchase was decking boards for the sides. He pulls it with his riding lawn mower to haul firewood, wood chips, mulch, soil, etc.

first time it was asked. Believing my purpose is to be happy, make money, or accumulate stuff seems like a surefire guarantee of an unhappy life. Personally, at the end of the day I want to feel that I've been productive. I don't necessarily have to complete all the items on my to-do list, but I at least want to check some of them off. But also, I don't want to be so focused on tasks that I don't enjoy the very reasons we chose this lifestyle in the first place. I want to remember to take the time to be still, to look, to listen, to feel and smell the air. I'm not just here to do, I'm here to be.

I'll be the first to admit that time stewardship isn't something I have figured out. Managing it is something I have yet to master.

> *"To keep things manageable, keep your schedule flexible and mentally allow for setbacks and delays. . . be willing to set things aside if you need to. . . accept that mistakes are a part of life."*
> *"Keeping Things Manageable,"*
> Prepper's Livestock Handbook *(pp. 193-194)*

I should learn to follow my own advice!

Time is often associated with money, and when I started planning this chapter I wondered if I should include it as a resource too. Then Dan pointed out that money is more of a tool than a resource. It's a human invention which is necessary for some aspects of modern life, but not as many as we are taught to think. Like other tools, it's something we acquire through a trade. In the case of money, we trade our time. In that regard, the challenge is keeping a balance, which I'll discuss that in chapter thirteen.

Notes

[1] Gabe Brown, "Sustainable Farming and Ranching in a Hotter, Drier Climate," 2017, https://youtu.be/clbhyVMtYc8.
[2] Elaine R. Ingham, *The Soil Biology Primer*, http://www.nrcs.usda.gov/wps/portal/nrcs/main/soils/health/biology/. Accessed March 30, 2019.
[3] Kris Nichols, "Does Glomalin Hold Your Farm Together?" *USDA-ARS-Northern Great Plains,* https://www.ars.usda.gov/ARSUserFiles/30640500/Glomalin/Glomalinbrochure.pdf. Accessed March 30, 2019.
[4] Jon Stika, *A Soil Owner's Manual* (CSIPP, 2016) 29.
[5] Joel Salatin, *Salad Bar Beef* (Swope, Virginia: Polyface, Inc., 1995) 192.
[6] Ibid., 219.
[7] Leigh Tate, *5 Acres & A Dream The Book* (Kikobian Books, 2013), 51.

Chapter 11

Discouraging Things

> *"Motivation is key to getting started and key to staying the course...
> But motivation by itself isn't enough. It gives us the enthusiasm to
> get started, and will often help us along the way, but there's more to
> it than that. We also need commitment."*
> "Introduction," 5 Acres & A Dream The Book *(p. 2)*

By our eigth summer on our homestead, my enthusiasm had gradually given way to discouragement. Gardens, pastures, permaculture hedgerows, and grain field all had problems, my bees had absconded, and nothing was doing as well as we thought it should. Garden yields were only fair, pasture weeds were more prolific than forage, and over the years I'd lost more than half of the trees and shrubs I'd planted. For some of it, I could blame the weather. Even so, I felt we should have it all figured out by now. We should be further along toward a thriving homestead. Honestly? It seemed like we'd taken at least as many steps backward as forward. Sometimes we wondered if we shouldn't give up. But to do what? I believe in this lifestyle with all my heart and soul. There is nowhere else to go.

Nothing challenges enthusiasm more quickly than things gone awry. It might be projects that don't turn out as anticipated or having to replan the day because of an unexpected change in the weather. It might be the endless learning curves that accompany homesteading. Figuring something out often seems slow because of the seasonal nature of the lifestyle: if something doesn't grow or produce well, I have to wait until next year's growing season to see if my improvement will work. Or, the simultaneous demands typical of homesteading might dampen enthusiasm. Many things need tending each day: critters, garden, cooking, preserving, etc. Progress on large projects is often slow because work on them must fit into routine chores, maintenance, repairs, and time off from one's job.

One source of frustration has been our never-ending list of uncompleted projects. Indeed, this can become overwhelming because we tend to focus on what isn't done, rather than what is.

> *"For every to-do list that I make, there are always things that don't get done. Usually, these are placed on the next list, then the next list, and sometimes the list after that. It took me a while to accept that I can't do it all and to stop scolding myself for it."*
> *"Keeping Things Manageable,"*
> Prepper's Livestock Handbook *(p. 194)*

Our master plan and prioritizing sessions help us make a good start on getting organized. We do our research and make a plan, but it's often difficult to follow through without interruptions—usually related to critters and/or fences. These are especially frustrating when they demand immediate attention, such as neighbors coming to tell us that our pigs have gotten out. Or a goat getting its horns stuck in a cattle panel. Or guinea fowl squawking in the middle of the road and blocking traffic.

Sometimes, things don't turn out as planned. One such project was my first permaculture hedgerow. It was exciting to start on this idea, because I love the concept of growing a sustainable, diverse, edible hedgerow. I spent a lot of time on research and compiled a list of compatible species, incorporating the existing blueberry, wild roses, pecan, and oak trees growing where the hedgerow would be.

I started my first year planting a canopy layer of pears, chestnuts, mulberry, hazelnuts, and Asian persimmons, choosing drought- and heat-tolerant varieties when I could. I planted aronia, horseradish, comfrey, echinacea, chicory, oregano, yarrow, and thyme. I amended planting holes generously with compost, then watered and mulched each planting well. I planned to plant more the following year.

A year later, less than half of what I originally planted had survived. Instead of expanding, the following year saw me replanting what had died. Identifying the causes wasn't difficult: regionally poor soil, a scorching summer, no rain, and chickens. Yes, chickens.

At the time, we still free-ranged our chickens. Because the hedgerow was fenced with cattle panels, the chickens had easy access to it. Had it been well established that wouldn't have been a problem. But the chickens zeroed in on the freshly planted trees and shrubs, scattering the mulch, and exposing the roots by scratching up the fresh soil. They killed more than one plant in my new food forest.

> *"The first thing I tried was to reroute them into another forage area. I closed all the gates and covered any chicken-sized openings. That stopped all but four persistent Buffs. I observed that most of the fence-hopping occurred at the cross members of the H-braces, so I used baling twine to create a barricade. This deterred them somewhat. They would make a terrible fuss when they jumped up and had to negotiate the twine, but it didn't stop them.*
> *"Chicken Tales: Chicken Wrangling,"* Critter Tales *(p. 65)*

My makeshift baling twine chicken deterrent was not a success.

Between destroying the hedgerow and eating much of the seed I planted for pasture improvement, we eventually decided that free-ranging was counterproductive for now. We expanded and improved the chicken yard (see chapter seven) and kept the chickens there.

Keeping the chickens out of the hedgerow has helped, but it is nowhere near how I envisioned it. Growth has been slow because of drought, and because perennial native grasses and weeds dominate the herbaceous and ground-cover layers. I occasionally cut these for hay, but they persist and push out much of what I plant. I like to think if the hedgerow was my only project it would be doing well. One dilemma of homesteading, however, is that one's time and energy are forever divided amongst simultaneous needs.

Other discouraging things have been harder to deal with, such as problems with animals. I always mean to do well by our critters, but when we were new to homesteading, we faced a hefty learning curve when it came to identifying and solving problems. The fact of the matter is, even when we're trying to do our best, bad things still happen. Those bad things include accidents, predator attacks, illness, and unexpected deaths.

Fortunately, we have never lost a critter from an accident. We've had goats, pigs, guinea fowl, and chickens escape under, over, or through fences, but none have been hit by a car or shot by a neighbor as trespassers. We have lost poultry to predators: hens to hawks, chicks to rats, and a Muscovy drake which disappeared overnight. We lost three hives of honybees because of skunks. Our list of local predators includes hawks, owls, snakes, rats, opossums, skunks, raccoons, foxes, roaming domestic dogs, and coyotes. We do our best to guard against this with proper fencing and housing, but also, we understand that some predation is inevitable. That doesn't make such losses any less discouraging, but that is the reality of keeping critters. Predation is part of the natural order of things.

I think what makes accidents and predator attacks so discouraging is that they seem preventable. Sometimes they are, but often they are beyond our control. Health problems, on the other hand, always seem within our control: I missed the clues, didn't understand what I was seeing, or lacked the knowledge to properly address the problem. Unfortunately, sometimes nothing helps.

> "The official diagnosis was Lyme Disease. This is caught from ticks, which was unexpected because I had been using flea and tick protection on both dogs. . . . Kris rallied a few days after he started the antibiotics, but several days later, he just seemed to give up;

> wouldn't move, wouldn't eat, and then he was gone. It was a real blow to both of us."
>
> *"Puppy Tales: Losing Kris,"* Critter Tales *(pp. 222-223)*

With livestock, there's been another challenge. Most of our local veterinarians specialize in pets and have little knowledge and experience with farm animals, especially goats.

> *"Even the vet was puzzled when I brought Chipper in. There was no diarrhea, no fever, no heavy parasite load. He just seemed to give up and then suddenly, death. I was devastated."*
>
> *"Goat Tales: Little Chipper & Old McGruff,"* Critter Tales *(p. 124)*

It took a long time to find a vet who understood goats, which meant I was often struggling to figure things out on my own.

> *"They were about three months old when, one morning, there was Dottie: weak, wobbly, and alarmingly thin. She was standing alone with a spaced-out look and trembling. What had happened? ... My best guess for Dottie was goat polio. This is actually caused by a vitamin B1 deficiency, not a virus or bacteria."*
>
> *"Goat Tales: Solved: Mystery of the Dying Kids,"* Critter Tales *(p. 162)*

I saved Dottie by administering vitamin B injections every six hours for the next week. The victory was bittersweet, however, because I realized her symptoms were similar to little Chipper's. Had I known about goat polio then, I might have been able to save him. Unfortunately, casualties are part of learning the hard way.

A promising buckling we called "The Grizz" always comes to mind when I think back on things I learned the hard way. What made his case especially disheartening was that I thought I was doing the right thing. By the time he went down, it was too late. When the vet arrived, she had a pretty good idea of what was going on—parasite overload. I was shocked because I had recently wormed him. The problem? Our parasites were now resistant to this class of wormer. She administered a wormer in the next chemical class, but it was too late. Unwilling to take food or water, he died a few days later.

My vet consoled me by telling me that goats are tough and show few symptoms until it's too late. When they finally go down, it's usually over quickly in spite of intervention. But I couldn't get over the fact that I was treating him with an ineffective product. What could I have done differently? Should I have had fecal exams for worm counts done regularly? Probably, but that isn't the thing one thinks about when believing one is doing the right thing.

It is said that experience is the best teacher, and I've taken my lessons to heart. One of the most important lessons is how a healthy animal looks and behaves. If something seems off, I immediately start looking for causes.

Health problems with animals seem like setbacks, but when they happen to the humans, illness and accidents can raise a worrisome uncertainty about the future. We faced that at the end of January 2018. Dan was outside working on the goat barn and I was inside doing computer work. I heard the back door crash open and Dan screaming. I jumped up and ran into the kitchen. He was in a panic, holding his left hand, blood everywhere. He had nearly severed a finger with the table saw. I grabbed a clean towel, wrapped his hand, and said, "we're going to the emergency room."

At the hospital emergency department, he was treated for shock, given an IV antibiotic, a tetanus shot, and had an x-ray taken. The flesh was badly torn, but the ER doctor told him he'd only lost a little bone on the tip of two fingers. The worst one was dislocated. The doctor numbed those fingers and realigned the dislocated one. He also told Dan that to save the fingers he would need to see a hand specialist. The doctor spoke with an orthopedic surgeon on the phone and made arrangements for Dan to meet him at the ER of a larger hospital. We were told to turn in Dan's paperwork at the desk when we got there and tell them Dr. So-n-So was expecting him.

It was late in the afternoon by then, so Dan wanted me to drop him off and go home to tend to the animals. He said he'd call when he was done. After checking in, he waited three hours at the second hospital's emergency department. When he finally asked about the surgeon he was supposed to meet, no one knew anything about it. Eventually, they took another x-ray. Why didn't they look at the one from the other hospital? Well, you never know, it might be blurry. Was Dr. So-n-So on his way? No answer, just another long wait. Finally, a first-year resident came in. She unwrapped the bandages, stitched up his fingers, and re-bandaged them, Dan was sent home with a prescription for pain pills. He was told to call and make an appointment with the orthopedic surgeon he was supposed to meet.

What happened to Dr. So-n-So? Could he not make the appointment? Had the hospital even contacted him? The questions were shrugged off. Not surprisingly, Dan was annoyed and questioned why he had bothered to come. We decided to make an appointment with our own doctor.

Dr. K took the time to show us the x-ray and explain what was going on. This was the first time we'd seen an x-ray of Dan's hand and were surprised by what we saw. Based on what we'd been told, we had assumed the bone loss was at the tip of the finger. Dr. K's x-ray showed that the saw blade had destroyed the knuckle. The flat ends of the bones that form the knuckle joint had been chiseled to points.

Regarding surgery, fusion was out because the ends of the bones were too badly damaged; there was nothing to fuse to. Letting it heal as-is was an option. The end of the finger was still very much alive and Dan had feeling and blood flow to the tip. Without a functioning joint, however, there would be limitations. The other option was amputating the end of the finger. The last thing Dan wanted was to lose the finger, so this was something of a blow. Fortunately, no decision had to be made on the spot. He had time to consider what to do.

All activity on the homestead reverted to basic functions. All projects were put on hold. I tended to daily chores, and we focused on wound care, pain management, and the emotional aftermath. I tried to steer conversation away from "what if" and "if only." It was enough to deal with our new daily routine.

Then the emergency medical bills started to come in. As an over-the-road truck driver, Dan didn't have sick time. Between the piling bills and no income, discouragement pressed in like a heavy weight. Because of our income, we qualified for financial aid but trying to get it was a nightmare.

The larger hospital's financial aid application required the occupations of patient and spouse, income, bank balances, value of home, and unpaid mortgage balance. We were able to get the application off promptly, although we contested the amount because he didn't go there for emergency care. They handled our case efficiently and most of the bill was eventually forgiven.

Our small local hospital, on the other hand, had requirements that seemed geared toward disqualifying people rather than helping them. I understood the need to evaluate income and assets, but for all members of the household—not just the responsible party—they wanted copies of drivers licenses or photo IDs, social security cards, 4 weeks of pay stubs, social security income verification, food stamp verification, self-employment verification, utility bills, car insurance statements, voter registrations, three months of bank statements for checking, savings, IRA/401k, and PayPal accounts, plus any other information on request.

PayPal accounts? Voter registration? Car insurance statements? Other information? What other information? And for all members of the household? So, if my great-grandmother lives with us and receives a small Social Security check each month to cover her personal needs, that would become part of our eligibility formula? Or if we had teenagers working to save for college, their personal savings is fair game for their parent's hospital bill? And what did our voter registration have to do with our financial situation? Where the larger hospital's application was logical, this one seemed absurd and overreaching. Were they really that suspicious of people's motives in requesting help? Were they trying to deter people from applying? Humiliate people for not being able to afford health insurance? And what were they going to do with all the irrelevant information once they got it? Based on their application, we didn't trust them enough to turn over all that personal information. As badly as we needed help, we balked. We used the last of our savings including our barn-building fund to pay the bill. The only consolation—besides sparing ourselves the humiliation of grovelling—was a ten-percent discount for paying the bill in full.

I'm guessing a number of you reading this are wondering why we would even need to do that. The U.S. now has universal health care, after all, so what was the problem? That misconception resulted in quite a bit of scolding from a few folks who don't understand that what we have isn't universal health care, it's mandatory health insurance. Unfortunately, there's no connection between the two because the health insurance industry is geared toward making profits, not helping people receive care. Considering how high deductibles have become, Dan and I would still have had to pay all his medical bills out-of-pocket.

Zeroing out our bank account left us wondering how we would make it. We still had miscellaneous medical bills dribbling in and still had to pay our mortgage, utilities, car insurance, phone, and internet bills. We still had to buy medication, wound care supplies, food, fuel, etc., and pay for follow-up doctor visits. It was hard not to feel stressed over the uncertainty of our situation. It was our online community who came to the rescue. Several people suggested I start a GoFundMe account, which I did. Dan and I have given money to folks in need this way, although it was awkward to ask for help ourselves. Between the generosity of people we'd never met and the forgiveness of part of our medical debt, we stayed afloat.

The next several months were tough. Dan decided not to have the amputation since the finger was still alive. He'd lost bone, muscle, and skin, so the wound continued to drain. Our daily routine revolved around wound care: flushing with Betadine, dressing with herbal salves, and fresh bandaging twice a day. Hot and cold treatments promoted good circulation to help with healing, as did good nutrition. I gave him bone broth for minerals, kefir and lacto-fermented foods for probiotics, detox tea and Bone, Flesh, and Cartilage tea and tincture for healing. Naproxen controlled the pain, but the hand was sensitive. It limited what he could do, but he pressed on with one hand as best he could. Between my two hands and his one, we were able to resume working on the goat barn.

One huge question in our minds was if and when he could go back to work. If driving a big truck was simply a matter of holding the steering wheel, no problem. That he could do, but things like strapping, tarping, and securing loads, carrying heavy objects, opening and closing stiff heavy container doors, connecting and disconnecting brake lines to the trailer, cranking the trailer's "landing gear" up and down for loading and unloading, even routine vehicle fluid checks; anything that required two hands he couldn't do. Initially, he was optimistic. He would say, "just a couple more weeks." But healing was much slower than we hoped, and eventually we had to look at other options.

The first option was Social Security Disability. Dan went down to the Social Security Administration building to apply. There, they discussed his options and told him that since he was 62, it would be faster and easier to apply for Social Security Retirement benefits. Even though the payments would be less than his driving income, that's what he decided to do.

From early on, one of our homesteading goals has been increasing our ability to live on less money. While most people are investing in their future retirement, we figured out a long time ago that this wasn't a reality for us. We've always lived paycheck to paycheck on a modest income because we didn't want to sacrifice family in pursuit of wealth. We knew it would mean a lean retirement, with less money than our modest working income had provided, but we'd been okay with that.

Then Dan had his accident. Long story short, "retirement" was now upon us several years earlier than we planned. We unexpectedly found ourselves face-to-face with a new normal. With it came a choice: we could rise to the challenge or give in to discouragement. The temptation is to worry: What if this happens? What if that happens? But I have to agree with something I once heard, that worrying is like paying interest on a debt not owed. Worrying leads to fretting and fretting leads to discontent and frustration. That's not a happy boat to be in.

In a sense, we'd been training for this all along. When Dan was between jobs we'd go weeks or sometimes months with little to no income. Each time, we learned to live a little more efficiently on what we had, rather than what we bought. We learned to ask, "Do we really need that thing?" and "Is there another way to accomplish the same goal?" We learned to look less at what we want and more at what we have. Each time, my frustration and anxiety over the loss of income decreased, and each time I learned to trust more in Providence.

While waiting for social security to kick in, we took a hard look at our old budget to determine how to adapt to a lower income level. Could we cut back on utilities? Phone? Internet? Food budget? Do we really need two vehicles? What is a necessity and what can we live without?

One area we saved money in was job-related expenses. Dan no longer needed food, supplies, and cash for miscellaneous expenses like laundry while on the road. He no longer needed to purchase safety and work gear. He no longer needed to travel back and forth to his big truck. That helped household spending quite a bit. He switched to a pay-as-you-go cell phone which was considerably cheaper. Then we cut back to one vehicle.

Because we had been working toward a less consumer-dependent lifestyle, we had already invested in tools and equipment that could help us toward that goal. We had gardening and food preservation tools and supplies. We had tools for hay and grain harvesting. We had carefully chosen equipment: a 45-horsepower farm tractor, the sawmill, and the chipper. We could continue to grow most of our own food and some of our livestock feed. We could produce our own lumber and wood chips. Every step we'd taken toward becoming more self-reliant added up. What could have been a discouraging time in our lives gave us a sense of freedom.

Except for Dan's permanently crooked finger, we rarely think back on that experience now. Even so, we still have to fight discouragement from time to time, especially when projects don't turn out as planned. Learning to call new projects "experiments" or "prototypes" helps with that. It relieves us of the mental pressure of having to get a thing right the first time around. Rainwater filtration, for example. When our "new" filter design turned out to be inadequate, the tendency was to think we got it wrong. But that gave us important information on what needed to be changed. By thinking of a project as an experiment, we aren't surprised when problems arise. In fact, we expect them and think of them as prompts toward a better solution.

A photo journal helps when we feel overwhelmed. In my case, it's my blog, which serves as the record of our homesteading journey. Looking through it from time to time reminds us of the changes and progress we've made. That helps keep the present in perspective. Another thing is to remind ourselves daily to take it one step at a time. Homesteading isn't a checklist, it's a lifestyle. It isn't a marathon, it's a journey.

Taking time to enjoy our critters and count our blessings are two ways to combat discouragement.

More recently, we've had to accept that we'll never get it all done. For every project we complete, a new idea presents itself. It may be the result of a problem-solving session, or a great idea we picked up from a website or YouTube. We evaluate each idea and add it to the project list if we think it will help us toward our goals. However, we've learned we can't hold ourselves responsible for accomplishing all those ideas. We keep working toward our goal, and we'll get as far as we get.

Discouraging things are a part of life. While we may not control our circumstances, we can control our attitude toward them. I can choose to be grumpy, or I can choose to be grateful. It all depends on what I think about; on where I put my mental focus. I'm not sure who said "just do the next thing," but it's excellent advice. Step by step, day by day, season by season, we just do the best we can. And when we look back, we see that we've come farther than we thought.

Chapter 12

DISTRACTIONS

"We are realizing that we must learn to ration our time and energy. We cannot afford wild bursts of enthusiasm which lead us down rabbit trails."
"Work Smarter, Not Harder,"
5 Acres & A Dream The Book *(p. 167)*

Discouraging things are immediately identifiable, but distractions are subtle. There are a lot of good ideas for homesteaders out there, and Dan and I make a lot of good plans. The problem is, there simply isn't enough time, energy, and money to accomplish it all. Having a master plan and knowing how to prioritize goals certainly helps. Sometimes, however, things pop up that don't necessarily conflict with our goals and plans, but become rabbit trails when injected with a healthy dose of enthusiasm.

This is probably a good place to point out that whether something is a distraction is subjective. Every time Dan and I research a topic we are presented with new information and different ways of doing things. The question we continually have to ask is how an idea would fit into our present workflow and whether or not it will help us reach our primary goal. We are constantly evaluating and reevaluating our projects and plans.

One example is our house. We chose a fixer-upper for the lower mortgage. It seemed more feasible to do the repairs and upgrades on a pay-as-we-go basis. When the time came to replace our dining room windows, we couldn't find good matches for replacements. So, we began to brainstorm. One idea was to take out not only the windows but the entire exterior dining room wall. In its place, Dan wanted to build a large bay window. However, this would involve tearing down the old chimney which used to service a coal stove in the dining room. And that would involve rebuilding the roof after the old chimney was removed. The more we discussed the idea, the more time-consuming the project became. We want our home to be aesthetically pleasing, but we also had to ask how elaborate we want to get.

Another example was building the goat barn. Since Dan has a sawmill, he was keenly interested in large timber construction. In the planning stages, he researched large timber techniques for building a large timber barn. We both loved the idea of an authentic style barn, but in the end, he used a variety of methods. His bottom line was a structurally sound building that meets our needs, not an authentic large timber barn.

I want to stress that our decisions for these two projects had nothing to do with right or wrong choices. Under other circumstances, we might have done things differently. We simply made choices that best helped us toward our primary objective of working within our budget to become a self-reliant homestead.

Our potential distractions aren't limited to building projects. Before Dan had his accident and was still over-the-road, an idea that we often discussed was buying a big

The gate between the milking room and goat quarters. The practical approach doesn't negate creativity.

truck and his becoming an owner-operator. On many levels, this makes sense. Being self-supporting dovetails nicely with self-sufficiency. Plus, money will always be a reality. As many times as we've discussed making money from home, being a trucker is what Dan knows. As a company driver, however, he is always at the mercy of his company—particularly his dispatcher—which means he can't control when he comes home, nor how long he's allowed to stay. Typical home-time for a truck driver is about a day and a half, no matter how long he's been on the road. With so much to do on the homestead, you can see why the job got frustrating. The idea of being his own boss, choosing his own loads, and scheduling his own time are all exceedingly attractive. Plus, owner-operators make more money which, for us, translated to not having to go out as often.

The argument to take this step was compelling, but we had to consider both sides of the coin. Start-up financing was one problem, which meant a loan to buy the truck. Collateral? We weren't willing to use the house for that. We also had to consider that all the maintenance and repairs would fall on Dan. Most of his home time would likely be spent tending to the truck. If a major breakdown occurs, well, it would be best to have savings set aside for big repairs. Besides, there are no guarantees as to how much he would earn. As much as he wanted to be his own boss, was everything else worth it? Deep down he knew it wasn't, so he looked at employment alternatives. He finally settled for a job driving weekends only. It paid less, but five days of home-time worked better with our goals than being gone most of the time.

My distraction has been my Kinder goats. Our goats play an important role on our homestead, but because they are registered, the role of being a dedicated breeder is always pulling at me. I love exploring pedigrees and studying genetics. I love inventing names for their paperwork. The challenge of working toward developing goats with good homesteading qualities is fun. Is there anything wrong with that? No! Is it incompatible with our goal of self-reliance? No! So what was the problem?

The problem is that it's too easy to allow my breeding goals to overshadow my homesteading goals. Our goats provide milk, meat, manure, brush control, and kids. The kids are necessary for milk and a sustainable supply of chevon. For year-around milk, I should have planned breeding so that all my does aren't pregnant and kidding at the same time. Instead, I found myself breeding so my kids would be close in age for potential buyers. That meant all does were bred about the same time, pregnant at the same time, and kidded around the same time. It also meant that they had to be dried up at the same time, hence, every year I'd have no

Not all distractions are bad, but they can cause an alteration in life course.

milk for several months. Then I'd have a mob of goat kids so that the barn seemed crowded and it was a strain on the pasture. I finally had to admit that I was putting my breeding goals ahead of my homesteading goals. I was letting my enthusiasm for something I enjoyed become a distraction.

Identifying distractions isn't easy because many of them are good ideas and good projects in their own right. It's easy to get excited about an idea or in trying to figure out how we can do it. It's also easy to become attached to an idea. Then, in our enthusiasm, it's easy to forget to evaluate how it supports our primary goal, or whether its cost in time and money is worth it. I'll take you through our evaluation process in the next chapter.

It seems everything we do points back to our primary goal, doesn't it? Of course, we could change that goal if we wanted to, but we don't. I suppose that's another way of saying that I don't regret the decisions we've made.

CHAPTER 13

Toward Keeping a Balance

"We must strive for a balance."
"Food Self-Sufficiency: Feeding Ourselves,"
5 Acres & A Dream The Book *(p. 81)*

Sometimes Dan and I feel like we're spinning our wheels. That's partially because of the many challenges I've been sharing with you in the previous chapters, especially when trial and error favors error rather than success. Another reason is because homesteaders have to balance many tasks: food production, food preservation, livestock management, resource management, mechanical maintenance and repair, building and fence maintenance, problem-solving, construction, etc. I've often thought if we had only one thing to focus on, it would be done well. As it is, we have many things to do. That means we must divide our time and energy well enough to keep our heads above water. It also means we have to let go of any preconceived notions of the perfect homestead.

I think this is a common experience for homesteaders. We're tackling a new way of life and have a lot to learn. In the beginning, we're busy trying to figure out how to translate the life we're used to into more natural ways and means. Unfortunately, it's nearly impossible to anticipate the learning curve. It seems like it ought to be a simple matter of swapping one way of doing things for another. Switching to homemade electricity, for example. We don't fully appreciate that the lifestyle we're accustomed to is only possible because of an abundant and easily available energy supply. To achieve success in a lifestyle transition, we have to learn how to change.

Hopefully, we're not just running on excitement but have taken time to define our goals and make a plan. Without these, life quickly becomes scattered. Even so, when working on simultaneous projects, it's easy to lose perspective. It's easy to feel like some things are being neglected for the sake of others. Somehow we must achieve some sense of balance.

What are we trying to balance? Often we think of balancing inputs and outputs, such as not having more grazing animals than the land can feed. For this chapter, I have identified three areas that have been challenging for Dan and me to balance: time and money, need and want, and work and rest. These aren't the only areas we struggle with, but they are the ones that keep rising to the surface. I can't say we've achieved complete success, but I can say we've made progress, and even a little progress helps.

Time and Money

I mentioned time and money in my chapter on evaluating priorities, but I've included the topic here because it tends to be a prominent imbalance in our lives. Part of the difficulty is because time and money are so closely tied in the modern lifestyle. Unless we inherit wealth, we have to use our time (as a job) to obtain money (as income).

Have you read Laura Ingalls Wilder's *Farmer Boy*? Do you remember the conversation between Almanzo and his father about the pumpkins?

> "Father asked: 'Almanzo, do you know what this is?'
> 'Half a dollar,' Almanzo answered.
> 'Yes, but do you know what half a dollar is? . . . It's work, son. That's what money is; it's hard work. . . . Say you have a seed potato in the spring, what do you do with it?'
> 'First you manure the field, and plow it. Then you harrow, and mark the ground. And plant the potatoes, and plow them, and hoe them. You plow and hoe them twice . . . Then you dig them and put them down cellar.'

> 'And if you get a good price son, how much do you get to show for all that work? How much do you get for half a bushel of potatoes?'
> 'Half a dollar,' Almanzo said.
> 'Yes,' said Father. 'That's what's in this half-dollar. The work that raised half a bushel of potatoes is in it.'"[1]

Several years ago, I wrote a blog post about growing our own goat feed. One comment I received was that this sounded like too much work. The thing is, it's work either way, whether I grow it myself or buy it. If a bag of feed is $15 and I make $15 an hour, then it "costs" me an hour of my time to purchase the feed (plus the fuel and time to go buy it). If I make $30 an hour, then it's only a half hour of my labor. The question then becomes, would I rather leave the homestead and work for someone else to pay for my feed, or stay at home and work for myself and my goats? While there's no right or wrong answer to that question, I think it's important to understand exactly what my choices mean in terms of time and money.

To work smarter not harder, we have to be realistic in our project choices and expectations. When evaluating an idea we've learned to ask:

Does the idea help us toward our goal of self-reliance?
How will it change our routine?
What kind of maintenance will it require?
How does it benefit us?
Will it decrease or increase our workload in the long run?
Will it make a difference a year from now? Five years? Ten years?
What are the alternatives?

These questions require trying to think a thing through beforehand. Because of them, we've changed our mind about a number of ideas, or at least put them on hold. A chicken tractor, for example. The benefits would be a more natural self-reliant diet for the chickens, also controlling the area they can graze and concentrating their bug eating, manuring, and manure scattering. But we'd have to move the tractor several times a day. How would that work with the corridor and gating system of our grazing rotation plan? Could we plug that task in as a regular chore? Or would it become an interruption of another task? Because we couldn't answer those questions satisfactorily, we set the idea aside for now.

The question of benefit is subjective because there are several ways in which we perceive it. Benefit might be functional, for example, if it meets a genuine need. It might be aesthetic if it beautifies or makes my environment more pleasant. Or a benefit might be personal. It might provide a sense of accomplishment or an increase in self-confidence.

Something we often debate is the purchase of tools and equipment. We've had to do enough things the hard way, i.e., by hand, to understand the importance of tools and equipment appropriate for our lifestyle. And we've made enough hasty spending choices to know that the immediate job at hand shouldn't be the only factor we consider. We also understand that higher tech isn't necessarily better tech. We're not technophobes, but neither are we technophiles. We're not so enamored with technology that we're willing to spend all of our time maintaining it and keeping it fed. To keep a balance on the homestead, we need sensible technology. We need technology that helps us work smarter not harder by getting the job done. Not a substitute of one kind of work for another.

To assess the condition of the old swimming pool, we had to decide between an expensive equipment rental or digging it out by hand.

Need and Want

Discerning the difference between need and want is often an extension of the time versus money debate. Not having abundant financial resources has often caused us to be reluctant to spend what we have. We find ourselves asking what we really need and what we don't.

"Having limited funds has turned out to be a blessing in disguise. True, we've had to do a lot more by hand, ending many a day with aching muscles reminding us that we're not getting any younger.

> *On the other hand, it's forced us to begin to think outside the box."*
> *"Work Smarter, Not Harder,"*
> *5 Acres & A Dream The Book (p. 168)*

Besides teaching us to be more creative in our problem solving, not having much money means we aren't as tempted by impulse spending. From experience, I find that impulse buys are ones that I usually come to regret. By having to approach our goals with limited funds, we've learned that it's okay to take time to figure out what we need, what we don't, and if there are alternatives.

Reaching a balance between need and want must address the question —how much is enough? There is a natural tendency in humans that pushes toward bigger and more. This may make sense in the economic growth model, but it makes zero sense in any ecological model. The longer output exceeds input, the worse environmental damage becomes. Learning how to live with less is an important life skill that helps us in resource stewardship.

> *"Our own entertainment consisted, in part, of reading books written by authors who lived in the late 19th and early 20th centuries. True, they had less, but they wanted less. There was a basic contentment and satisfaction with life that is missing in society today."*
> *"The Establishment Phase,"*
> *5 Acres & A Dream The Book (p. 48)*

The key to finding a balance between need and want is contentment. Unfortunately, contentment is an emotional skill that isn't easy to learn. For me, having limited monetary resources helps teach me how to be content. When the gotta-have-it bug starts biting, I can either become indigent with my financial situation or I can accept it. Once I disengage emotionally from wanting something, I almost always find that I'm doing just fine without it.

Emotional attachment to an idea is another obstacle in trying to discern need versus want. It might be an attachment to the project itself, or to a particular way the project ought to be done. Even when it logically makes sense to not pursue a particular idea, we sometimes can't let go.

Emotional attachment has likely undermined many a joint decision. "A" wants this, "B" wants that, and nothing gets decided because the two can't get past having their own way. At the time it seems so important to be "right," but as Dan likes to say, "Who's gonna notice ten years from now? Who's gonna care?" If Dan and I can't agree on something, we set it aside. Our to-do list is certainly long enough to find something else to focus on. After some time has passed, it's easier to discern whether the idea is a legitimate need or a want. The interesting thing is, if we later go forward with it, we usually come up with better ideas than we had at first.

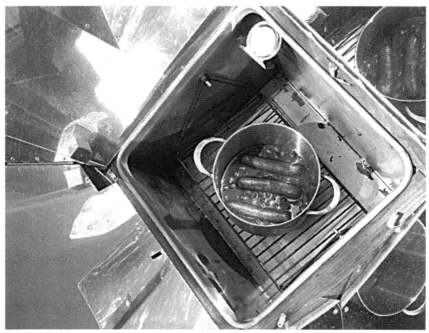

We work steadily toward our goals and ideal lifestyle, but contentment comes easier if we aren't dogmatic about it.

Work and Rest

In chapter ten I addressed time as a resource and wrote of time pressure, i.e., the mental and emotional drive to achieve a goal within an expected time frame. With our goal of self-sufficiency, the time frame is "before we're too old to accomplish it all." That's not very specific, but it creates a sense of time pressure nonetheless. The problem is that time pressure can be a relentless and stressful driving force. To stay the course and avoid burn-out, we need to balance that drive with rest.

It's important to understand what I mean by a balance between work and rest. I don't mean a balance of:
- work and play
- work and fun
- work and alternative activities

Play, fun, and alternative activities can be important, but they easily can be as physically, mentally, and emotionally strenuous as work. Play and fun often include competitive activities which key us up rather than calm us down. The same is true for alternative activities. Spending one's day off running errands and catching up on life maintenance chores such as shopping, laundry, cleaning, is still work! As can be filling a day set aside for worship with fellowship activities. These things can be just as busy and non-restful as hard physical and mental work.

Rest should be more than a distraction. It should refresh us physically, mentally, emotionally, and spiritually.

Physically refers to the physical body. There's only so much our muscles, joints, and nerve synapses can take. Constant activity will wear the body down. That kind of weariness and stress makes us more susceptible to injury and illness.

Mentally refers to busyness of the mind. Think of how many battles we fight in our minds! Some of us are constantly trying to combat problems or figure something out. In homesteading, there are ongoing problems and challenges to work out. We can become mentally stressed when we can't put it down. Constant mental activity can interfere with sleep, which interferes with our physical rest. Mental rest is just as important as physical rest.

Emotional rest is the reprieve from ongoing emotional ups and downs, from emotional turmoil. So many things trigger emotional responses nowadays, anything from politics to being stuck behind someone driving five miles under the speed limit. In homesteading, the culprit is often fretting over not having enough time to accomplish everything. Yet, it's wearying to be continually frustrated, annoyed, angry, or on edge.

And what is spiritual fatigue? It's a weariness of spirit. Ask someone who suffers from chronic pain. They are tired to the core of their soul from always hurting, always being in pain, from no relief. Eventually it wears down one's will to live. In homesteading, spiritual fatigue leads to burnout.

The answer for these is rest. Rest should be like a cool breeze on a hot summer day, a cool drink of fresh water for a parching thirst. Rest should be refreshing. The problem is that we develop a lifestyle habit of busyness. We always feel like we have to be doing something, so we gravitate toward distractions, which are only substitute activities.

Obviously, the balance between work and rest is not a fifty-fifty proposition. To take a more realistic approach, I'd like to point out two time-honored patterns for rest.

The first is the most ancient of rest models. There, we see a six-to-one work-to-rest ratio. Six days of work, one day of rest. Traditionally, this was called the Sabbath. Literally, it means to cease, end, rest.

It's interesting that originally, the Sabbath never referred to a day of worship, although that's what it's come to mean in modern times. For a Believer, worship is an important part of life, but true worship requires taking our minds off of ourselves and our needs, and focusing on gratefulness toward something greater than ourselves. I can't do this if my mind is racing with needs, worries, and complaints. I can't truly worship if my mind isn't at rest.

> *"Most of the advice I've read about avoiding homestead burnout recommends taking an annual vacation. The problem is that for most homesteaders a real vacation isn't practical. Even if we can manage to find someone knowledgeable and willing enough to care for our livestock for a week or two, it usually isn't financially feasible. Instead of taking vacations, Dan and I keep a day of rest. One day every week is set aside to rest, relax, and reset our minds, bodies, and spirits."*
>
> "Keeping Things Manageable,"
> Prepper's Livestock Handbook *(p. 194)*

On Dan's and my day of rest, we set aside everything except essential critter chores. We don't get up as early, our meals are more leisurely, and we set aside time for Scripture study and prayer. We take time for leisurely walks around our property and visiting with our critters. We take time to enjoy all the things there are to see. We make a point of being thankful for what we have. Having a set day of rest helps me get through busier times of the week. When things seem hectic, I know that a break is imminent; a reset from the busyness, cares, and concerns of the week before. That helps keep me going.

We find the other pattern of rest in the seasonal cycle of the agrarian year: sow, grow, harvest, and rest. Indeed, now that we've pretty much worked our way through the establishment phase of our homesteading, Dan and I find our lifestyle and its accompanying activities follow these four seasons. Spring is for planting, summer for growing, autumn for harvest and food preservation, and winter is the season of the hearth, the season of rest. These seasons aren't strictly defined and can vary depending upon where one lives. But they represent the natural ebb and flow of the agrarian lifestyle.

It's the seasonal rhythm of our lifestyle that helps us maintain the more practical aspects of balance, for example, livestock numbers. Too many critters put stress on the land, our feed sources, and our time. Too many critters and things are no longer manageable.

> *"For Dan and me, autumn is the time we take stock of our livestock numbers and availability of feed for the winter. Our goal is to keep a balance. Having to overwinter too many animals tips that balance."*
> "Eggs, Milk, and Meat," Prepper's Livestock Handbook *(p. 142)*

The natural pattern of livestock numbers is seasonal. Young arrive in the spring and grow when the forage is thriving. Autumn is the traditional time for selling the extras and harvesting meat. In terms of balance, it makes sense to have fewer critters when forage is dormant.

The seasonal cycle also helps keep a balance with project load, especially during planning sessions because many routine jobs have set times. Seasonal requirements such as planting, harvesting, and cleaning the chimneys have priority at specific times of the year. Building, maintenance, and repair fit well into other seasons, such as summer or winter.

Living with the rhythm of the seasonal cycles helps me keep perspective during busy times. When I'm spending day after day in the kitchen with my cutting board and a hot steamy canner, I remind myself that this is just a season. The intensity of food preservation will pass, and I'll soon have time for other pursuits. That helps keep me going too.

So how are we doing in these three areas: time and money, need and want, work and rest? How well do we maintain a balance? It isn't easy because balance isn't intuitive to human nature.

> *"The concept of working smarter, not harder, is one that we are realizing is vital to our success at homesteading. We are realizing that we must learn to ration our time and energy. We cannot afford wild bursts of enthusiasm which lead us down rabbit trails. Nor do we need to start projects that will increase our work load in the long run. There are things we need to accomplish if we are working toward self-sufficiency, but we need to do so intelligently."*
> "Work Smarter, Not Harder,"
> 5 Acres & A Dream The Book *(pp. 167-168)*

At some undefined point, Dan and I realized we'll never get it all done. That there will never be enough time, money, and energy to do all the wonderful things we want to do. That was a crisis point at which we had to examine how we were thinking about our goals. Are they a checklist that we must complete? Is not doing so a failure? If we can't do it all, then what's the point of trying? What are we doing here, anyway? What is our purpose? More on that next.

Notes

[1] Laura Ingalls Wilder, *Farmer Boy*, (New York: HarperCollins, 1953) 182-184.

Conclusion

A Sense of Purpose

> *"The primary goal for our homestead is sustainability. We want our homestead to achieve and maintain a balance. We believe it is an essential part of stewardship."*
> *"Goat Tales: It Was Time for Elvis to Go,"* Critter Tales *(p. 138)*

This quote points out a shift in thinking that has gradually occurred over the years. Even now, as I write these chapters, I often refer back to our goal of self-sufficiency. I don't think we realized it in the beginning, but I now see that goal as being linear, while our lifestyle has become more cyclic. That's partially a result of living in the seasons, but it's also a reflection of what our homesteading years have taught us. It's less about becoming self-reliant and more about being good stewards.

Philosophers say there are fundamental questions that each person asks and answers throughout their life. These questions may not be asked consciously, but usually come through an innate curiosity and inner pondering. The answers are rarely found through personal conclusions, but by observing the people around us: what they say, how they react, what they believe, how they behave, what emotions they display, how they make choices, what their goals are, and what's important to them.

One of those questions is "why am I here?" Or "what is my purpose in life?" Perhaps a century ago and more, people found purpose in their work, in what they did for a living. Making a contribution to society (when was the last time you heard that phrase?) provided that sense of purpose and worth. That doesn't seem to be the case today. People rarely talk about a sense of personal fulfillment found in their careers. Some people love their jobs, but how many do you know who don't? Who desire to do something else?

I think one reason is because most jobs today are tiny parts in a larger system. We do the same thing day after day with no sense of being connected to that system's ultimate purpose. We may not even know what that purpose is, other than to make money for someone else.

Since the industrial revolution, our world has been on a fast track away from the natural creation and toward a world shaped by human hands. We humans take pride in our technology and believe it can solve all problems. We believe technology and the science that created it can save the world.

Do I think technology is bad? Absolutely not. But I see a steep cost to handing all aspects of our lives over to technologists. That cost is an ever-increasing disconnect between the world they are creating and the natural creation. Many people no longer know where their food and water come from. Or why all animals aren't pets. Or why we need farmers. It's now possible to live one's entire life without ever seeing a tree or walking barefoot in grass.

So if people no longer find purpose in their work, to what do they look instead? If I were to ask you what your purpose is, how would you answer? I'll guess that common answers might be: to be happy, have fun, become wealthy, love others, be the best version of ourselves, do God's will, save the earth. Some will say there is no purpose to life.

What does a sense of purpose have to do with homesteading? To me, everything. I believe humans are supposed to be a part of the natural earth system, not create a new world system. Homesteading is how Dan and I seek to do this. We believe this is how we're supposed to live. We are meant to be participants who individually make a difference, not bystanders who hand everything over to politicians. Let's face it, farming is the oldest profession in the Book. God: "Adam, you and Eve are to grow food, tend the animals, and take good care of the earth. It's your responsibility now." No matter how humans try to rationalize abandoning that model, I personally can't get away from it.

This gradual revelation has had a profound impact on how we view everything: homesteading, economics, politics, the environment, etc. It has slowly changed our primary goal from self-reliance to becoming a functional part of our homestead ecosystem. Hopefully, next time around, I'll have much to tell you about that.

Homestead Recipes

Homegrown recipes for self-reliance

Fiesta Cornbread	193
Oven-Fried Okra	194
Probiotic Ice Cream	195
Nutrient-Rich Bone Broth	196
Apple Pectin	197
Spicy Fig Jam	199
Heavenly Chèvre Cheesecake	200
Chèvre (a soft goat cheese)	200

Fiesta Cornbread

A meal in a skillet, baked the traditional Southern cornbread way.

For the Cornbread

Preheat a 10-inch cast iron skillet in the oven at 425°F (220°C) with ¼ cup of your favorite fat or oil.

Mix:

1½ cup cornmeal (I used homegrown, home ground)
½ cup unbleached white flour (makes the cornbread less crumbly)
1 tsp salt
2 tsp baking soda
1 tsp chili powder
1 egg
1 cup buttermilk, whey, yogurt, or kefir

Pour batter into the hot skillet, return to the oven, and bake till half done (about 10 minutes). While the cornbread is baking, prepare the topping.

Topping:

In another hot skillet brown:
    Handful of chopped onion
    Handful of chopped green pepper
    Ground meat or meat substitute
    Salt, pepper, and chili powder to taste

Have ready:
    Canned chili beans (or your favorite cooked dry bean)
    Grated cheese
    Chopped fresh tomato

Remove half-baked cornbread from the oven. Top first with beans, then browned meat mixture, then the cheese, lastly the tomato. Finish baking (about 10 more minutes or until inserted knife comes out clean).

Slice into wedges and serve immediately with a salad for a complete meal.

Recipe notes:

Baking cornbread in a preheated cast iron skillet makes a delicious cornbread with a delightfully crispy crust. If you are using a regular baking pan instead of a skillet, add the melted fat or oil to the batter.

Substitute canned or leftover chili for the beans, tomatoes, and meat.

I like to preheat the beans (or chili) before spooning them onto the half-baked cornbread. I find cold beans slows the baking.

You may have noticed that there is no sweetening in this cornbread. This is because we don't care for sweet cornbread unless we're eating it drizzled with honey for dessert.

Oven-Fried Okra

Just as tasty as deep-fried okra, but uses less oil.

Ingredients:

Fresh okra, sliced
Olive oil
Cornmeal
Seasonings to taste (we like salt, pepper, and garlic)
Either more olive oil, or oil or melted fat of choice

In a bowl, mix sliced okra and enough olive oil to coat well. Sprinkle with cornmeal and seasonings, stirring and adding cornmeal until the "strings" are gone. Generously grease a cookie sheet with more olive oil, or whatever oil or fat you like. Spread the okra on the baking sheet and bake at 425°F (220°C), stirring frequently, for about 20 - 25 minutes (or until golden brown). Enjoy hot out of the oven.

Probiotic Ice Cream

Make a deliciously tangy-sweet ice cream loaded with probiotic goodness. The proportions are for a small, freezer bowl ice cream maker, but you can adjust them to fill whatever ice cream maker you have.

Ingredients:

    1 cup cream
    1½ cups crushed fruit (strawberries, peaches, etc.)
    dash sea salt
    ½ cup unbleached sugar (or equivalent sweetener)
    3 egg yolks
    1½ cups kefir (or yogurt)

Mix cream, sugar, fruit, and salt in a saucepan. Heat until almost simmering. In a blender, beat the egg yolks. Slowly add the hot fruit mixture and mix well. Chill thoroughly. Before pouring into the ice cream maker, add the kefir or yogurt and blend until smooth. Freeze according to ice cream maker manufacturer instructions. Makes a generous quart plus.

Recipe Notes:

Heat kills probiotics, but they remain alive in this ice cream because they are added cold.

You can use fresh, canned (and drained), or frozen (thawed) fruit.

You can substitute any kind of milk for the cream.

Adjust sweetener to taste. For tart fruits, you may need a little more.

For lemon ice cream, add ½ cup lemon juice, 1 tbsp lemon zest, and increase sugar to 1 cup.

Nutrient-Rich Bone Broth

The champion of soup and gravy stocks. Loaded with collagen, calcium, and other minerals dissolved from the bones by an acid in the cooking water.

To Make:

>Meat bones popped into a bag in the freezer after eating the meat.
>Onion, large, chopped
>Celery stalks with leaves, several chopped
>Carrots, several chopped
>Water
>½ cup vinegar (any kind), lemon juice, or wine (any kind)

If there is still quite a bit of meat on the bones, I simmer the bones alone for about a day. After the pot cools, I remove the bits of meat from the bones and use in soup, enchiladas, or a meat pie. The bones and broth are returned to the stock pot and the veggies and vinegar are added. Do not change the water. Simmer slowly for two or three days. The longer the simmer, the more minerals are extracted. Cool, skim the fat, and strain.

When the broth is chilled, it should congeal from the collagen.

To Can:

Reheat broth to boiling. Fill canning jars leaving ½-inch headroom. Add ½ teaspoon canning salt per pint, and 1 teaspoon per quart. Wipe jar rims and secure lids. Pressure can 20 minutes for pints and 25 minutes for quarts at 10 pounds pressure (adjusted for your altitude).

To Use:

Use in soups and gravies, or as a substitute for water when roasting meat or cooking rice or grains.

Apple Pectin

Make jams and jellies without commercial pectin and without huge amounts of sugar.

Ingredients

Crabapples or green (not ripe) apples of any variety
Water

Apples do not have to be peeled or cored. Chop them and place in a heavy pot. Cover with water and simmer over medium heat. Cook the apples down to a thin apple sauce. Cool and pour into a jelly bag. Hang the bag and allow the liquid to drain into a large bowl.

Squeeze or press out as much of the liquid as you can. This is the pectin, but it should be tested first to see how well it will gel (gel test below).

If it passes the gel test, it is ready to use, to freeze, or to can for future use. If it fails the gel test, simmer it down to concentrate the liquid more. Then try the gel test again.

Yield will depend on the number of apples you used and how much the liquid is cooked down.

Cooked apple pulp in a jelly bag. I feed the dry pulp to the chickens.

Gel Test:

¼ cup isopropyl (common rubbing) alcohol, 70% or higher
1 tsp of your liquid pectin

Put the alcohol into a small jar or glass and add the pectin. Allow to sit one minute, then try to fish the pectin out with a fork. You should be able to lift out a soft glob of gelled pectin as in the photo below. If it doesn't gel, cook down the apple liquid to further concentrate the pectin, then retest.

Successful apple pectin gel test.

To Can:

Pour pectin into half-pint jars and process in boiling water bath for 15 minutes. May also be frozen.

To Use for Jam or Jelly:

Per 1 cup of mashed fruit or juice:
    ¼ cup pectin
    ¼ cup sugar

Bring to a slow boil, stirring constantly, until the jam sticks to a spoon.

Recipe Note: You can adjust sugar to taste.

Spicy Fig Jam

August is our month to harvest figs, and this is one of my favorite fig jams. Perfect for my homemade apple pectin.

Ingredients:

    6 cups mashed figs
    1½ cups sugar (¼ C sugar per 1 C mashed fruit)
    1½ cups homemade pectin (¼ C pectin per 1 C mashed fruit)
    ¼ cup plus 1 tablespoon lemon juice (1 tbsp per 1 C fruit)
    1½ tsp cinnamon
    1 tsp ground ginger
    ½ tsp ground cloves

Stir together all ingredients and bring to a boil. Simmer until it sticks to a spoon. Will thicken more as it cools.

To can:

Spoon into sterilized jars leaving ¼ inch headroom. Apply lids. Process in boiling water bath for 10 minutes.

Heavenly Chèvre Cheesecake

Chèvre is creamy soft goat milk cheese that is easy to make but expensive to buy. You can substitute ricotta or cream cheese for the chèvre.

To Make the Chèvre:

- 1 gallon fresh goat milk
- ¼ cup fresh kefir (can use cultured buttermilk or cultured whey)
- $1/8$ of a regular rennet dose
- 1 tablespoon good quality salt

Pour milk into a heavy-bottom pot and stir in the kefir. Slowly warm to 90°F (32°C). Add the rennet and stir well. Cover the pot and let sit until the curd sinks to the bottom of the pot (12 to 24 hours). Ladle into a cheesecloth lined colander, cover, and drain for 6 hours. Then mix in the salt, tie up the cheesecloth like a bag, and let drain for another hour or two. After that it's ready to eat! One gallon of milk yields the amount needed for this recipe.

Cheesecake Filling

- 3 cups chèvre (can use ricotta or cream cheese)
- 3 eggs
- ½ cup sugar
- dash salt
- 1 cup heavy cream, whipped
- 1 tsp vanilla (or whatever flavoring you wish)
- prepared crumb or cheesecake crust

Beat eggs and sugar until light. Add salt and vanilla and mix well. Add chèvre and blend. Fold in the whipped cream. Pour into crust and bake at 350°F (180°C) for one hour. Turn off oven, crack the door, and let the cheesecake cool inside the oven for another hour.

Yield: enough for one 12-inch cheesecake pan or two 9-inch pie pans.

Appendices

APPENDIX A

Resources

In addition to the bibliography, here you'll find information for the various books, websites, equipment, tools, and products I've mentioned or shown throughout the pages of this book. My listing them here is not an endorsement, rather, it's to help you start your own research on items you found of interest. I will mention that we've been happy these products. The page numbers indicate where the item is pictured or mentioned.

Books & eBooks

The Art of Natural Cheesemaking by David Asher (p. 76), Chelsea Green Publishing, 2015.
Beekeeping With a Smile by Fedor Lazutin (p. 70), New Society Publishers, 2020.
Five Acres and Independence by M. G. Kains (p. 135), Dover Publications, 1973.
Hands-On Agronomy by Neal Kinsey (p. 146), Acres U.S.A., 2009.
Handy Farm Devices and How to Make Them by Rolfe Cobleigh (p. 135), Skyhorse Publishing, 2007. Also published online at http://journeytoforever.org/farm_library/device/devicesToC.html. Free.
Homemade Contrivances and How to Make Them by Rolfe Cobleigh (p. 135), Skyhorse Publishing, 2007.
How To Preserve Eggs: Freezing, Pickling, Dehydrating, Larding, Water Glassing, & More by Leigh Tate (p. 75), Kikobian Books, 2014. http://kikobian.com/little_series_preserve_eggs.html. Free eBook.
Prepper's Total Grid Failure Handbook by Alan and Arlene Fiebig (p. 111, 113), Ulysses Press, 2017.
Preserving Food Without Freezing or Canning by The Gardeners & Farmers of Terre Vivante (p. 76), Chelsea Green Publishing, 1999.
Sepp Holzer's Permaculture (p. 61), Chelsea Green Publishing, 2011.

Websites

American Guinea Hogs (p. 33), https://guineahogs.org.
Bone, Flesh, & Cartilage tea and tincture (p. 171), https://www.herballegacy.com/Bone_Flesh_&_Cartilage.html.
Craigslist (international listings), https://www.craigslist.org/about/sites.

"How Many Solar Panels & Batteries Do You Need For Your Off-Grid System?" (p. 114), https://preparednessadvice.com/solar/many-solar-panels-batteries-power-grid-system/.

Kinder goats (p. 177), http://kindergoatbreeders.com.

Layens natural beekeeping (p. 70), https://www.beeculture.com/the-layens-hive/.

Library for Solar and Windpower Learning, https://www.altestore.com/howto/.

SmartMix cover crop calculator, https://smartmix.greencoverseed.com.

USDA Web Soil Survey (p. 62), https://websoilsurvey.sc.egov.usda.gov/app/HomePage.htm.

Warré natural beekeeping (p. 70), https://www.warre.biobees.com.

How-Tos

African Keyhole Garden (p. 142), https://www.5acresandadream.com/2020/05/african-keyhole-garden.html.

Beet syrup and sugar (p. 71), https://www.5acresandadream.com/2018/05/may-harvest-sugar-beets.html.

Cheesemaking:
- Domiati (p. 76), https://www.5acresandadream.com/2019/02/domiati.html.
- Farmer's Cheese (p. 77), https://www.5acresandadream.com/2017/10/new-adventures-in-cheese-making.html.
- Feta in Herbed Olive Oil (pp. 76-77), https://www.5acresandadream.com/2017/11/a-tale-of-two-fetas.html.
- Gjetost (Norwegian Whey Cheese, p. 78), https://www.5acresandadream.com/2017/10/gjetost-norwegian-goat-whey-cheese.html.
- Halloumi (p. 76), https://www.5acresandadream.com/2019/08/cheesemaking-halloumi.html.
- Mozzarella (pp. 76-77), https://www.5acresandadream.com/2015/06/mozzarella-making-revisited.html.
- Paneer (and how to fry it. p. 77), https://www.5acresandadream.com/2017/12/fried-cheese.html.
- Whey Ricotta (p. 78), https://www.5acresandadream.com/2015/08/a-simple-ricotta-cheese.html.

Convert a chest freezer to a fridge (pp. 127-128), https://www.5acresandadream.com/2020/03/chest-freezer-to-fridge-conversion.html.

Dry-Pack Vacuum Canning (p. 73), https://www.5acresandadream.com/2017/02/dry-pack-vacuum-canning.html.

Feed chopper from a yard chipper (p. 69), https://www.5acresandadream.com/2015/08/diy-goat-feed-experiment-1.html.

Grain thresher from a yard chipper (p. 69), https://www.5acresandadream.com/2017/07/wheat-processing.html.

Olla (p. 142), https://www.5acresandadream.com/2020/07/conserving-water-in-garden-olla.html.

Tools, equipment, & other products mentioned

I have to say I'm rather dismayed at how prices have jumped since we bought these items. I recommend that you shop around for the best price. I did not include solar panels, batteries, cables, and hardware for our solar project because everything was customized for our particular system and set-up. For solar panels, I recommend checking Craigslist for bargains on contractors' leftovers.

12-volt DC box fan (p. 113), https://amzn.to/2XqfW2e.

Carpet sweeper (p. 105), https://amzn.to/2MQzWWU.

Charge controller, Flexmax 60 (pp. 122 & 125), http://www.outbackpower.com/products/charge-controllers/flexmax-60-80.

Charge controller remote monitor, Outback Mate2 (p. 125), https://www.altestore.com/store/inverters/inverter-accessories/remote-controls/outback-mate-remote-monitor-control-p857/.

Cover crop chart (p. 151), https://www.ars.usda.gov/ARSUserFiles/30640000/PDF/CCCv1-2.pdf.

Dough whisk (p. 104), https://amzn.to/2zotWBI.

Ecofan (p. 111), https://ecofan.caframobrands.com.

Garden cart (p. 152), https://www.northerntool.com/shop/tools/product_200641807_200641807.

Grain mills (p. 104)
    Country Living Grain Mill, http://www.countrylivinggrainmills.com.
    WonderMill, http://www.thewondermill.com.

Hayfeeder plans (p. 41), https://www.premier1supplies.com/img/instruction/127.pdf.

Hay rake (p. 152), https://www.onescytherevolution.com/hay-rakes.html.

Kill-A-Watt electric usage monitor (p. 114), https://www.homedepot.com/p/Kill-A-Watt-Electricity-Monitor-P4400/202196386.

Laundry plunger (p. 139), https://www.breathingwasher.com.

Laundry wash tub stand (p. 139), apparently no longer available.

Layens beehives (p. 70, free plans), http://layenshive.com/how-to-build/hive-frame-swarm-trap.shtml.

Mosquito Dunks (p. 137), https://www.summitchemical.com/mosquito/mosquito-dunks/

Rain catchment tank, 1550 gallon (p. 132), https://www.tractorsupply.com/tsc/product/liquid-storage-tank-1550-gal.

Remote Temperature Sensor (p. 125), https://amzn.to/2zrvcnJ.

Rennet, natural, non-GMO, https://www.walcoren.com/en/index.html.

Sawmill (p. 26), Harbor Freight Central Machinery, https://www.harborfreight.com/saw-mill-with-301cc-gas-engine-62366.html.

Screen mesh colanders (p. 88), https://amzn.to/2LZrOD9.

Scythe (p. 106), https://www.onescytherevolution.com/quick-order.html.

Seed for pasture, hay, cover crops (p. 151), https://greencoverseed.com/.

Shade cloth (pp. 67-68), https://shadeclothstore.com.

Soil amendments and organic fertilizers (p. 147), https://www.7springsfarm.com.

Solar attic fan (p. 107, 109), https://amzn.to/2TVjNnh.

Solar battery box DC cooling fan (p 124), https://amzn.to/2XwnMHF.

Solar cooker plans, (p. 108), https://solarcooking.fandom.com/wiki/Category:Solar_cooker_plans.

Solar oven (p. 108 and 184), https://www.sunoven.com.

Solar shed light (p. 112), https://amzn.to/3cZ7Bci.

Tattler reusable canning lids (p. 72), https://reusablecanninglids.com.

Thermostat controller for freezer to fridge conversion (p. 127), https://amzn.to/2zhAp10.

Utility carts
    Foldable kitchen cart (p. 109), https://amzn.to/36xWncI.
    Solar recharging station cart (p. 101), https://amzn.to/2M2RYF5.

Vacuum sealing (dry pack vacuum canning, p. 73),
    Pump-N-Seal Food Saver Vacuum Sealer, https://pump-n-seal.com.
    FoodSaver canning jar attachments, https://amzn.to/3bTpI2h.

Vortex blender (p. 105), https://gsioutdoors.com/vortex-blender.html.

Warré beehive free plans (p. 70), https://warre.biobees.com/warre_hive_plans_imperial.pdf.

Wash tubs (p. 139), https://www.tractorsupply.com/tsc/product/behrens-hot-dipped-steel-tub-square.

Wringer (p. 139), https://wisementrading.com/laundry-supply/wringer/wringer-hand-clothes-black-2/.

Wood chipper, WoodMaxx WM8M (p. 158), https://www.woodmaxx.com/WM_8M_Mechanical_PTO_Wood_Chipper_p/wm-8m.htm.

Water glass (liquid sodium silicate, p. 75), https://www.lehmans.com/product/water-glass-liquid-sodium-silicate.

Appendix B

Pasture Rotation: 3 Models

When I began researching grazing rotation, I was looking for a set of specific rules. What I discovered was that pasture rotation advocates don't necessarily agree on any one plan. As I took notes and made lists, the information eventually seemed to sort itself out into three models, each with different goals and recommendations.

Health of the Forage

Preventing animals from overgrazing and killing forage is a concern for livestock owners. Grazing animals will eat what they like best first, and leave the less palatable growth to eventually take over. Overstocking is usually identified as the cause, so measures are taken to keep stock numbers low. Rotating grazing for the sake of the forage gives it time to recover, so that desirable plants are able to thrive.

The focus of this model is the height of the forage. There are specific guidelines for different types of forage, but in general:

Don't graze forage height below four inches.
Allow 20 to 30 day for forage recovery.
Allow grazing again when forage is 8 to 10 inches tall.

Health of the Animals

The particular concern here is internal parasites, especially worms. Pasture rotation is part of "integrated parasite management" (IPM). IPM includes use of wormers based on test results (fecal or FAMACHA i.e. level of anemia) rather than a schedule, developing the animals' immune system through culling and breeding for resistance, good sanitation practices, forage that promotes parasite resistance, and pasture rotation.

IPM guidelines for pasture rotation:

Don't graze forage below a couple of inches.
Multispecies graze to "vacuum" the fields. Cattle aren't susceptible
    to the same parasites as goats and sheep.
Don't graze any paddock more than three to five days.

Don't graze when forage is wet (parasite larvae need moisture to climb plants so they can be ingested).
Don't overstock paddocks.
Allow pasture to rest. Recommendations vary between 21 to 65 days, depending on the particular parasite problem.
Some sources are now recommending mowing down to about an inch or two in height to allow sun and air circulation to dry the soil and kill larvae.

Health of the Soil

Properly managed, intensive rotational grazing (mob grazing) builds soil health by sequestering carbon from plant residue through animal impact. This is accomplished by:

Broad forage diversity
High stocking density (250 - 500 cows per acre).
Rotating stock once or more per day
    (rule of thumb - graze 50%, leave 50%).
Long forage rest (5 to 6 months).

Why does it work? Because when livestock are concentrated in small areas, they will either eat it or trample it. However, they are moved before they overgraze (graze 50%, leave 50%). The trampled forage begins to decay which in turn feeds soil organisms, helps retain soil moisture, sequesters carbon, and builds the soil. The long rest period allows forage to recover fully before being grazed again. The key is monitoring forage and knowing when to move them. The challenge is that rotation and rest cycles are highly variable, depending on the season, weather, type, and condition of the forage.

For Further Reading:

"Grazing Height Determines the Health of Your Forages," https://onpasture.com/2017/07/10/grazing-height-determines-the-health-of-your-forages/.
"Integrated Parasite Management For Livestock," https://attra.ncat.org/viewhtml/?id=258#ref.
"Mob Grazing 101," https://hereford.org/static/files/0111_MobGrazing.pdf

Appendix C

Polyculture Forage, Hay, and Cover Crop Lists

People sometimes ask me what to plant for their livestock. I'm no expert, but I have researched the topic extensively and can offer some suggestions. Mind you, these lists are not exhaustive! But hopefully, they will be a start in helping you plan a diverse polyculture for pasture, hay, or cover crops.

I plant as many of these as I can, with some from each group. Native grasslands grow hundreds of species, so there's no limit on what I try to plant. Long-leafed grasses provide energy and roughage, nitrogen-fixing legumes provide protein, and broadleave forbs provide vitamins and minerals. Together, they provide a rich diversity of plant life necessary for healthy soil. Some things do better in certain conditions than others, but with a wide range of seeds, something always comes up.

Key: w (warm season), c (cool season)
a (annual), p (perennial), b (biennial)
ap (cool climate annuals, warm climate perennials)

Grasses

Annual ryegrass: *Lollium multiflorum*, c, a, erosion control, soil builder, weed suppressor, drought tolerant.

Barley: *Hordeum vulgare*, c, a, erosion control, soil builder, weed suppressor, nitrogen recycler, good grazing. Grain crop, good for biomass.

Big Bluestem: *Andropogon gerardii*, w, p, native grass, erosion control, good forage and hay.

Brome: *Bromus*, c, p, soil builder, erosion control, nitrogen scavenger, weed suppressor, drought, heat, and salinity tolerant. Grazing and hay.

Eastern gamagrass: *Tripsacum dactyloides*, w, p, a native prairie grass. Fast growing, erosion control, good grazing, loosens subsoil, drought and flood resistant. Used for riparian buffers.

INDIANGRASS: *Sorghastrum nutans*, w, p, a native grass. Heat and drought tolerant, erosion control, attracts birds.

MILLET: Varieties listed below are are good for grazing and hay.
BROWNTOP: *Panicum ramosum*, w, a, erosion control, suppresses root-knot nematode, drought tolerant. Often grown for wildlife.

GERMAN (FOXTAIL): *Setaria italica*, w, a, soil builder, erosion control, drought tolerant, nitrogen scavenger, weed suppressor.

JAPANESE: *Echinochloa esculenta*, w, a, soil builder, erosion control, drought and heat tolerant, nitrogen scavenger, weed suppressor. Excellent regrowth.

PEARL: *Pennisetum glaucum*, w, a, soil builder, erosion control, rapid warm weather growth, good regrowth, nitrogen scavenger, weed suppressor, high yield.

OATS: *Avena sativa*, c, a, quick growth, good grazing, erosion control, weed suppressor, green manure. Grain crop. Also good for biomass.

ORCHARD GRASS: *Dactylis glomerata L.*, c, p, erosion control, soil builder, weed suppressor, nitrogen scavenger, good for hay and grazing, good regrowth. Easy to establish but not winter hardy.

RYE: *Secale cereale*, c, a, quick growth, drought tolerant, erosion control, soil builder, weed suppressor, loosens topsoil. Grain crop, biomass. Very cold hardy.

SORGHUM (forage types), SORGHUM-SUDAN (hybrid, also called sudex), and SUDANGRASS: *Sorghum bicolor*, w, a, quick growth, erosion control, soil builder, good forage, hay, drought and heat tolerant, weed and nematode suppressor, subsoil loosener, biomass.

SWITCHGRASS: *Panicum virgatum*, p, w, native grass, cold tolerant, erosion control, biomass, hay and forage. Also grown for wildlife.

TEFF: *Eragrostis tef*, w, a, fast growing, erosion control, soil builder, nitrogen scavenger, weed suppressor, good grazing and hay, heat and drought tolerant, tolerates multiple cuttings. Grain crop in Africa.

TIMOTHY: *Phleum pratense*, p, c, erosion control, soil builder, nitrogen scavenger, weed suppressor, loosens topsoil, biomass, prefers moist soil.

TRITICALE: *Triticale hexaploide* (wheat/rye hybrid), c, a, drought tolerant, soil builder, erosion control, weed suppressor, biomass. Grain crop.

WHEAT: *Triticum aestivum*, c, a, erosion control, weed suppressor, adds organic matter to the soil. Grain crop.

NOTE: Grain grasses can be harvested for hay when the cereal grains are forming but the plant is still green and leafy. Makes a highly palatable hay.

LEGUMES

ALFALFA (LUCERNE): *Medicago sativa*, c, p, nitrogen fixer, soil builder, drought tolerant, subsoil improver, good grazing, green manure, hay.

BIRDSFOOT TREFOIL: *Lotus corniculatus*, w, p, nitrogen fixer, erosion control, ground cover, good forage, wildlife, tolerates wet soil.

CHICKPEAS (GARBANZO BEANS): *Cicer arietinum*, a, w, nitrogen fixer, soil builder, subsoil aerator, grazing, tolerates dry conditions.

CLOVER: A sampling of the many types available.
  ARROWLEAF: *Trifolium vesiculosum*, c, a, nitrogen fixer, soil builder, drought tolerant, good for grazing, hay, and wildlife. Reseeds well.

  BALANSA: *Trifolium michelianum*, c, a, nitrogen fixer, soil builder, cold and wet condition tolerant, good grazing and hay, biomass, erosion control, excellent reseeding rate in no-till systems.

  BERSEEM: *Trifolium alexandrinum*, c, w, a, nitrogen fixer, weed suppressor, erosion control, green manure, good grazing, wildlife food plots, quick growth, cold tolerant.

  CRIMSON: *Trifolium incarnatum*, c, w, a, nitrogen fixer, soil builder, erosion control, forage, subsoil aerator, biomass, nectary plant.

  RED: *Trifolium pratense*, c, a, b, nitrogen fixer, soil builder, forage, weed suppressor, attracts pollinators and beneficial insects.

Sweet (Yellow Blossom): *Melilotus officinalis*, w, b, nitrogen fixer, soil builder, erosion control, subsoil aerator, salinity tolerant, green manure, grazing, biomass, nectary plant, attracts wildlife.

White (Ladino, Dutch): *Trifolium repens*, c, p, nitrogen fixer, soil builder, erosion control, green manure, grazing, living mulch, tolerates heavy traffic, attracts beneficial insects.

Cowpeas: *Vigna unguiculata*, w, a, nitrogen fixer, soil builder, cover crop, grazing, hay.

Lentils: *Lens culinaris*, c, a, nitrogen fixer, grazing, hay, weed suppressor, green manure, drought tolerant but not tolerant of wet soils.

Sainfoin: *Onobrychis viciifolia*, c. p, nitrogen fixer, soil builder, subsoil aerator, weed suppressor, erosion control, good grazing and hay. Doesn't tolerate heavy grazing.

Sericea lespedeza: *Lespedeza cuneata*, w, p, nitrogen fixer, hay, drought tolerant, good grazing in dry weather, anthelmintic (anti-parasite) in small ruminants. Can be invasive. Classified as a noxious weed in some states.

Sunn hemp: *Crotalaria juncea*, w, a, nitrogen fixer, soil builder, green manure, cover crop, weed suppressor, sequesters nitrogen, heat tolerant, fairly drought tolerant, nematode control, biomass. Doesn't tolerate heavy grazing. Best suited to tropical and subtropical regions.

Vetches:

Common: *Vicia sativa*, c, a, nitrogen fixer, green manure, cover crop, erosion control, grazing, hay.

Hairy: *Vicia villosa*, c, a, nitrogen fixer, green manure, erosion control, topsoil conditioner, cover crop, nectary, weed suppressor, grazing, hay.

Woollypod: *Vicia villosa ssp. dasycarpa*, c, a, nitrogen fixer, soil builder, weed suppressor, erosion control, biomass, nectary.

Winter peas (field peas): *Pisum sativum*, c, a, nitrogen fixer, green manure, weed suppressor, forage, hay, biomass.

Forbs

Buckwheat: *Fagopyrum esculentum*, w, a, quick cover crop, topsoil loosener, weed suppressor, attracts pollinators and beneficial insects, not recommended for grazing.

Chicory: *Cichorium intybus*, w, p, soil builder, drought tolerant, mineral accumulator, good grazing, attracts beneficial insects, anthelmintic (anti-parasitic) in small ruminants.

Collards: *Brassica oleracea*, c, a, soil nutrient scavenger, very cold tolerant, high forage & biomass production, nematode control, reduces compaction, tolerates heavy grazing.

Daikon: *Raphanus Sativus*, c, a, fast growing, erosion control, weed suppressor, reduces soil compaction, biomass, increases water infiltration, good grazing, wildlife food plots. Roots can be chopped for winter feed.

Kale: *Brassica oleracea*, c, a, winter hardy, cover crop, good grazing, highly palatable and nutritious, erosion control, breaks hardpan, builds subsoil.

Mangels (Forage, fodder beets): *Beta vulgaris*, use non-GMO seed only. c, a, b, all parts edible, breaks up subsoil, fairly drought tolerant, high nutritive value, high yield, good grazing for livestock and deer. Roots can be chopped for winter feed.

Okra: *Abelmoschus esculentus*, w, a, fast growing, soil builder, drought resistant, breaks subsoil compaction, rich in vitamins, winter snow catch.

Phacelia: *Phacelia tanacetifolia*, c, a, b, fast growing, builds soil, reduces soil compaction, sequesters nitrogen, weed suppressor, cold and shade tolerant, good grazing and hay, attracts pollinators, self-sowing.

Plantain (narrowleaf): *Plantago lanceolata*, w, p, erosion control, soil building, weed suppressor, nitrogen scavenger, good grazing, mineral accumulator, anthelmintic (anti-parasitic).

Purple Coneflower: *Echinacea purpurea*, w, p, soil builder, drought tolerant, attracts pollinators, medicinal.

Radish (forage): *Raphanus Sativus*, c, a. quick growth, soil builder, erosion control, nitrogen scavenger, weed suppressor, breaks up hardpan, good grazing.

Rape (canola): *Brassica napus*, c, a, erosion control. soil builder, nitrogen scavenger, weed suppressor, good grazing, quick growin

Small burnet: *Sanguisorba minor*, c, w, p, hardy, evergreen, drought tolerant, erosion control, excellent forage, attracts pollinators, used in wildlife plots. Doesn't tolerate wet conditions.

Sugar beets: *Beta vulgaris*, use non-GMO seed only. c, a, hardy, quick growing, highly palatable, soil conditioner, tolerant of high traffic, salinity tolerant, not drought tolerant. Roots can be chopped for winter feed.

Sunflower (black oil seed): *Helianthus annuus*, w, a, erosion control, soil builder, nitrogen scavenger, weed suppressor, potential for grazing and hay, drought tolerant, salinity tolerant, attracts pollinators and beneficial insects. Seeds can be harvested and fed in the shell.

Turnip (purple top): *Brassica rapa, c, a, b,* erosion control. soil builder, nitrogen scavenger, weed suppressor, good grazing and hay, quick growing.

Also see Appendix C in 5 *Acres & A Dream The Book* for a list of herbs and "weeds" that contribute to a diverse polyculture pasture.

Sources:

Cover Crops Canada, http://covercrops.ca.
Managing Cover Crops Profitably by SARE (Sustainable Agriculture Research and Education, 2010). Free PDF edition available at https://www.sare.org/publications/covercrops/covercrops.pdf.
"Soil Health Resource Guide," Sixth Edition (Green Cover Seed, 2020). Available at https://greencoverseed.com.

Bibliography

Atkins, Robert Wayne, "Five Different Shelf Life Studies: Two on Canned Food and Three on Dry Food," *Grandpappy's Official Website*, 2007/2010. https://grandpappy.org/hshelffo.htm.

--- *Grandpappy's Recipes for Hard Times.* Grandpappy, Inc., 2011.

Brown, Gabe, "Sustainable Farming and Ranching in a Hotter, Drier Climate," 2017. https://youtu.be/clbhyVMtYc8.

Cover Crops Canada, accessed May 25, 2020. http://covercrops.ca.

Dixon, Amanda, "How Much Is the Average Electric Bill?," *SmartAsset*, July 9, 2019. https://smartasset.com/personal-finance/how-much-is-the-average-electric-bill.

Ferguson, Gillian, "Milk as a Management Tool for Virus Diseases," *Greenhouse Grower Notes*, Ontario Ministry of Agriculture, Food and Rural Affairs. November 1, 2005. http://www.omafra.gov.on.ca/english/crops/hort/news/grower/2005/11gno5a1.htm.

Fukuoka, Masanobu, *The One-Straw Revolution.* NYRB, New York, 1978.

Gordon, Kindra, "Mob Grazing 101," *Hereford World*, January 2011.

Ingham, Elaine R., The Soil Biology Primer. http://www.nrcs.usda.gov/wps/portal/nrcs/main/soils/health/biology/. Accessed March 30, 2019.

Jorgustin. Ken, "Temperature Versus Food Storage Shelf Life," *Modern Survival Blog.* January 5, 2015, https://modernsurvivalblog.com/survival-kitchen/temperature-versus-food-storage-shelf-life/.

Kidwell, Boyd, "Mob Grazing: High-density stocking builds profits back into the cattle business." *Angus Beef Bulletin*, March 2010.

Managing Cover Crops Profitably. Sustainable Agriculture Research and Education, 2010.

Nichols, Kris, "Does Glomalin Hold Your Farm Together?" USDA-ARS-*Northern Great Plains*. https://www.ars.usda.gov/ARSUserFiles/30640500/Glomalin/Glomalinbrochure.pdf. Accessed March 30, 2019.

Salatin, Joel, *Salad Bar Beef.* Swope, Virginia: Polyface, Inc., 1995.

Savory, Allan, & Butterfield, Jody, *Holistic Management: A Commonsense Revolution to Restore Our Environment.* Island Press, 2016.

Sloane, Eric, *The Seasons of America Past.* Mineola, New York: Dover Publications, 2005.

Smart, Paige, "Grazing Height Determines the Health of Your Forages," *On Pasture.* Modified June 26, 2017. https://onpasture.com/2017/07/10/grazing-height-determines-the-health-of-your-forages/.

Soil Health Resource Guide, Sixth Edition. Green Cover Seed, 2020.

Stika, Jon, *A Soil Owner's Manual*. CSIPP, 2016.

Tate, Leigh, *5 Acres & A Dream The Book*. Kikobian Books, 2013.

--- *Critter Tales: What my homestead critters have taught me about themselves, their world, and how to be a part of it*. Kikobian Books, 2015.

---"Double Digging for Rainwater Collection," *5 Acres & A Dream The Blog*, April 10, 2017. https://www.5acresandadream.com/2017/04/double-digging-for-rainwater-collection.html.

--- "Dry-Pack Vacuum Canning," *5 Acres & A Dream The Blog*. February 7, 2017. https://www.5acresandadream.com/2017/02/dry-pack-vacuum-canning.html.

--- *How To Bake Without Baking Powder: modern and historical alternatives for light and tasty baked goods*. Kikobian Books, 2016.

--- *How To Preserve Eggs: freezing, pickling, dehydrating, larding, water glassing, & more*. Kikobian Books, 2014.

--- *Prepper's Livestock Handbook*. Ulysses Press, 2018.

"Temperature Coefficient (Q10) Calculator," *PhysiologyWeb*. Updated August 28, 2015. https://www.physiologyweb.com/calculators/q10_calculator.html.

Wells, Ann, "Integrated Parasite Management For Livestock," *ATTRA – Sustainable Agriculture Program*. Updated December 15, 2014. https://attra.ncat.org/viewhtml/?id=258#ref.

Wilder, Laura Ingalls, *Farmer Boy*. New York: HarperCollins, 1953.

INDEX

Page numbers in *italics* indicate photos, illustrations, quotes, or captions.

Numbers

5 Acres & A Dream The Book, 1, *3*, *3*, *8*, *9*, 10, *11*, *13*, *14*, *16*, *18*, *21*, *31*, *32*, *53*, *55*, *57*, 61, *65*, 68, *72*, *76*, *83*, *85*, 89, *101*, *104*, *106*, 111, *116*, *131*, 131, 137, 138, 143, 146, 155, *159*, *160*, *163*, *175*, *179*, *183*, *187*
6-volt batteries, *122*, 125
12-volt
    adapter plug, *113*
    batteries, 113-115, 125
    battery bank, 122
    box fan, 113, *113*, 205
110-volt, 110
220-volt, 110

A

accident(s),
    Dan's, 19, 168, 171, 176
    livestock, 166
acidic soil, 90
activities
    alternative, 184, 185
    church, 185
    lifestyle, 186
    seasonal, 16, 17
adapter plug
    kit, *120*
    plug, 113
African keyhole garden, 142, 204
aggregates, soil, 61, 148
agrarian
    calendar, 17
    lifestyle, 186
    rhythm, 18
    year, 18, 186
agricultural lime, 90
agriculture, sustainable, 13, 14
air conditioning (AC), 82, 102, 106
albumin, 78

algae, 137
alkaline, 90
almanacs, 17
alternating current (AC), 115, 116
alternative, 137, 177
    activities, 184, 185
    food storage, 80, 116
    solutions, 8, *23*, 25-27, 55, 181, 183
    tools, 104, *104*, *105*
amaranth, 87
American Guinea Hogs, 33, 98, 203
amperage (amps), 120, 122, 125
amp-hours (AH), 113-115, 125, 126
amputate, 169
analysis, 29, 52
    soil, 147
angry, 185
animals, 155, 168, 190
    balance, 146, 187
    dairy, 47
    farm, 167
    feeding, *85*-99
    genetic potential, 88
    grass-fed model, 95
    grazing, 49, 92, 99, 180, 207
    immune system, 207
    meat, 46-47
    immune system, 207
    meat, 46-47
    problems with, 166-168
    production model of, 89
    stewardship of, 14, 15
    rainwater for, 132, 137
    relationship with, 99
annoyed, 168, 185
annual
    crops, 151
    forage, 89-91, 209-214
    grasses, 91, 209-211
    soil testing, 147
    vacation, 186

(annual continued)
    vegetables, 150
    weather patterns, 61
    woody plants, 91
antibiotics, 81, 166, 168
apple
    cider vinegar, 139
    pectin, 197-198, 199
    trees, 45
appliances, 82, 104, 110, 113-115, 126, 129, 130
    energy-efficient, 116
appropriate, 14
    tools and equipment, 54, 144, 182
array (solar), 119-122, 124
Art of Natural Cheesemaking, 76, 203
aseasonal, 69
Asher, David, 76, 203
assets, 169
    tangible, 19
Atkins, Robert Wayne, 71
attachment
    emotional, 183
    FoodSaver jar sealer, 73, 206
    tractor, 145
attic fan, 83, 107, *107*, *109*, 111, 206
attitude, 173
Austrian winter peas, 89
authentic timber barn, 176
autumn (fall), 76, 186, 187
    growing season, 67
    harvest, 69, 187
    season, 16, 186
auxiliary fridge, 82, *128*

B
Bacillus thuringiensis israelensis (BTI), 137
bacteria, decomposing, 148
    nitrogen-fixing, 148
    pathogenic, 167
    putrefying, 138
    soil, 56, 148
    unwanted, 76
baking
    eggs for, 75
    soda, 77, 139, 140

balance, 15, 27, 84, *146*, 162, 179-188, 189
    bank/mortgage, 169
    need and want, 182-184
    time and money, 180-182
    work and rest, 184-187
barefoot, 190
barley, 90, 209
barn, 33, 111
    bedding, *91*
    bench, *157*
    buck, 51, 52
    building, 34, 36, *37-44*
    cleanings, 97, 151, 158
    coal, 3, *5*
    downspout, 135
    goat, 3, 22, 25, 26, 27, *28*, *32*, 34, 35, 55, 90, 132, 155, *156*, 168, 171, 176, 178
    lime, 90
    litter, 91
    new, 5, 51
    original, 5,
    quilt, 3, *44*
    roof, 132, *133*
    savings fund, 155, 170
    tanks, 134, *136*
    timber, 176
barnyard, 34, *46*
barter, 33
battery/batteries, 111, 116, 127, 204
    6-volt, *122*, 125
    12-volt, 122, 125
    bank, 114, 115, 126, 129, 130
    box, *121*, *122*, *124*, 126
    charging, *125*
    cranking (starting), 115, 125
    deep cycle, 113, *113*, 114, 115, 125
    golf cart, 125
    lead acid, 125, 126
    lithium, 125
    marine, 125
    nickle iron, 125
    off-gassing, 121
    replacement, 126
    RV, 125
    sealed, 109, 125
    wiring, 122

bay leaves, 81
bay window, 3, 176
bedding
    barn, *90, 91*
    deep litter, 90, 91
beds
    borders, 156
    filtration, 137, *137*
    garden, 12, 14, 55, 59, 146, *149*
    grazing, 97, 98, 137, *141*
    hoop house, 67
    hügelkultur, 61, 63-65, 131, 151
    in-ground, 59
    raised, 59, *60*, 61, 67
    strawberry, 59
beehives, 32, 70, 166,
    Layens, 70, 205
    Warré, 70, 206
benefit, 12, 67, 73, 82, 112, 144, *146*
    of soil carbon, 147-149
    question of, 181
    retirement, 171
Bermuda grass, 58, 90
Betadine, 170
big truck, 171, 172
biochar, 27, 153
biodegradable, *138*, 138, 140
biorhythm, 160
Black Australorps, 69
blackwater, 138
bleach, 137, 138
    chlorine, 140
    oxy, 138-139
    recipe for homemade, 139
blender, 69, 75, 84
    vortex manual, 105, 206
blessing(s), 173, 182
bloom (eggs), 75
Bon Ami, 140
bone
    broth, 81, 149, 196
    fusion, 169
Bone, Flesh, & Cartilage, 171, 203
boom pole, 145
borax, 140
boron, 138, 140
box fan(s), 107
    12-volt, 113, *113*, 205

brainstorm, 52, 176
bread machine, *110*
Brigham Young University, 80
brine, 76, 79
brine-cured cheese, 76
broadcasting seed, 91, 151, 152
Brown, Gabe, 147, 149, 152
brunost (Norwegian brown cheese),
    78, 78, 204
buck(s), *33, 47, 50, 167*
    barn, 51, 52
    shelter, 49, *50*
bucket and chain, 69
buckets as filters, 135
buckwheat, 213
budget, 102, 106, 115-117, 126, 147, 172,
    176
bulk, *88*, 147, 151
    grains, 81
bumper car approach, 22
burn-out, 184
bus bar, *122*, 125
busyness, 159, 185, 186
butchering, 33, 33, 70, 99

C

cables (solar), *120-122*, 125
calcium carbonate, 90, 140
calendar, 16, 17
canning, 73, 79, 82, 129
    dry pack vacuum, 73, *73-74*, 204,
        206
    garden, 150
    heat from, *109*
    lids, reusable, *72*, 206
    scraps, 99
    water, 140
canopy layer, 49, 164
carbon, 147-150, 153, 157
    dioxide, 147
    layer (mulch), 148 152
    liquid, 147
    nitrogen ratio, 149
    sequestering, 92, 149, 208
carbonic acid, 149
carpet sweeper, 105
carport, *139*

casein, 78
cash crops, 149, 151
cattle panels, 54, *94*, 164, 165
Cecil sandy loam, 62
ceiling fans, 107, 111
chainsaw
    crosscut chain, 155
    mini-mill, 155
    ripping chain, 155
challenge(s), 1, 18, 55, 56, 84, 164, 171
    food production, 58, 65, 68, 138, *142*, 146
    food storage, 73, 79, 80
    fridge conversion, 127-129
    keeping a balance, 162, 179, 185
    livestock, 46, 167, 177, 208
charcoal, *109*
charge controller, 113, 115, 116, *122*, *125*, 125-127, 129, 205
checklist, 12, 24, 54, 55, 172, 188
cheese, 74, 76
    aging, 76
    brined, 76
    brunost, 78
    as protein, 69
    cave, 76, 77
    Chèvre, 200
    curing, 76
    Domiati, 76, 204
    Farmers, 77, 204
    feta, 76, 77, 204
    fresh, 76, 77
    Halloumi, 76, 204
    mozzarella, 76, 204
    paneer, 77, 204
    Queso Blanco, 77
    ricotta, 78, 204
    starter cultures, 76,
    Mediterranean, 76
    northern European, 76
    storage, 25, 76, *77*, 77, 81
    traditional, 76
    whey, 78, 204
cheesemaking, 33, 76
    how-tos, 204
    whey from, 33, 77
chest freezer, 82, *82*

conversion, 116, 127, 204
chest fridge, refrigerator, 127, *128*
chestnut trees, 45, 54, 164
chevon, 177
chickens, *33*, *57*, 70, 73, 77, 80, 83, 88
    and compost, *97*, 97, 158, 159,
    and ducks, 96
    coop, 3, 4, *7*, *21*, 34, 34, 55, 69
    feeding, 73, *87*, 95, *96*, 96-98, 181, 197
    free-ranging, 54, 55, 165-166
    grazing beds, 97-98, 137
    happy, 97
    manure, 90
    numbers, 95-96
    problems, 54, 55, 164, 165, 166
    Black Australorp, 69
    tractor, 181
    watering, 137
    yard, 4, 34, 47, 48, 54, 96, 157
chickweed, *66*
chimney, 22, 176, 187
chipmunk, 136, *136*
chipper
    conversion, *69*, 205
    industrial, 157
    PTO, 145, 157, *158*, 172, 206
    yard, 69
chlorine bleach, 140
choice(s), 32, 148, 189
    beehives, 70
    budget, 182
    food, *65*
    greywater, 140
lifestyle, 9, 10, 171
    project, 8, 26, 176
    time vs. money, 26, 181
chores
    daily, 18, 24, 169
    essential, 186
    maintenance, 185
    morning, 50, 160
    routine, 164
    seasonal, 19, 54
circuit
    breakers, 117, 122, 125, 127
    closed, 49

circumstances, 173, 176
clay soil, 61-63, 148
cleaners,
    greywater safe, 140
    problems with, 138
clean-out plug, 133, *133*
climate, 34, 55, 60, 61, 65, 66, 71, 76, 81, 83
clothes dryer, 69, *110*
clothesline, *110*
clover, 68, *89*, 90, 221-222
commitment, 55, *163*
companion planting, 68, 151
compost, 33, 63, 91, 97, 137, 146, 149, 151, 165
    bins (piles), 34, 96, 97, 141, 157, *159*
    keyhole, 142
    tea, 27
    wood chip, 158
compressor, 82, 129
concentrates, 86, 95
conduit, 47, *120*, *122*
connectors, 115, 117, 125
consumer
    lifestyle, 172
    system, 9-10, 12
consumerism, 8,
content, 8, 183
contentment, 20, *183*, 183
contract, solar, 112
control, 2, 14, 88, 160, 166, 173, 177
    brush, 177
    chickens, 181
    erosion, 209-214
    fridge temperature, *127*, 206
    insect, 90
    nematode, 213
    weed, 12, 92
convenience, 82, 104
conventional farming, 12
conversion
    freezer to fridge, 116, 129, 204, 206
    yard chipper to thresher, 69, 205
cool climate pasture, 209
cool weather crops, 67
cooling fan, *124*, 206
cooperative extension, 147
corn, commercial, 86
corn stalks, 63
corral, goat, 36
cost
    air conditioning, 106
    analyzing, 126
    batteries, 114, 125
        replacement, 126
    calculating, 26
    laundry, 138
    livestock, 95
    shipping & handling, 147
    solar system, 111-112, 116, 117, 125
    time & money, 178, 181
    technology, 190
    tractor, 145
    windows, 102
counter-cultural, 7
counterproductive, 95, 166
Country Living grain mill, *104*, 205
cover cropping, 46, 149
cover crops, 149, 151, 152, 204
COVID-19, 2
cowpeas, 86, *87*, 149, 212
Craigslist, 26, 113, 116, 155, 203, 205
cranking (starting) batteries, 115, 125
crawlspace, *121*
Creation, 8, 15, 189, 190
credit, 27, 126
Critter Tales, 1, *33*, 34, 69, 70, 86, *96*, 97, 99, *146*, 165-167, *189*
crock-pot, 78, 79, *110*
crosscut chain, 155
culture, 3, 8, 20
cultures (starter), 76, 200
curiosity, 115, 189
curtains, 107
cushaw, 87
cycles
    production, 69
    rest, 208
    seasonal, 18, 90, 187
Cynodon dactylon, 58

D
dairy
    animals, 47, 94
    goats, 69

(dairy continued)
    products, 80
daylight savings time, 160
debt, 170, 171
decision-making, 9, 13, *31*
deep litter, 90
dehydrated foods, 25, 73, *74, 75*, 79, 80
detergents, 138, *138*, 139
devil grass, 58
diagram, battery wiring, 122
diagram, solar systems, 119
diet
    chickens, 98, 181
    homegrown, 83
    normalcy, 84
    seasonal, 17, 83
    simplified, 2, 12
digital device, *10*
direct current (DC), 113, 115, 116
disc, 145
disconnect, 8
discontent, 171
distractions, 175-178, 185
diversity
    diet, 83
    forage, 89, 90, 93, 151, 208, 209
DNA, 88
"do without," 8
dogmatically, 184
dogs, 34, 166, *166*
dogwoods, *45*
Domiati, 76, 204
dominant species, 8
door,
    barn, *136*
    crawlspace, *121*
    fridge, 116
    hayloft, 3, *44*
    hinges, *118*
    house, 102, 140
dough whisk, *104,* 205
drainage
    pipe, 135
    problems, 51
    water, 62, 141
drought, 59, 89, 138, 166
drought resistant (tolerant), 89, 90, 164, 209-214
dry pack canning, 73, *73-74*, 206
dry well, 51
duck house, *141*
duck pond (pool), 137
ducks, 95, *96*, 98
    Muscovies, 48, 70, 96

E

earth grounding, 122, 123
earthworks, 61
earthworms, 56, *97*, 98
ebb and flow, 186, 205
ecofan, 111, *111*
economic
    collapse, 10
    growth, 20, 182
    system, 9, 19
economics, 190
    sustainable, 13
economists, 143
ecosystem
    grassland, 99
    homestead, 7, 56, 190
edge slab uses, 155
egg(s), 13, 15, 17, 69, 76, 95, 98
    beater, *104*
    cooking, 83, 108
    dehydrated, *75*
    duck, *141*
    feeding to pigs, 98
    insect, 90
    powdered, 75
    preserving, 75
    production, 88, *96*
    storage, 25, 75, 81, *128*
    surplus, 80
electric
    alternative tools, 104, *105*
    bill, 101, 106, 107, 126, 129
    fence, *45*, 47, *49, 52*, 92, 111
    heat, 101, 111
    netting, 47
    skillet, 110
    stove, 110, 129
    washing machine, 139
electrical grid, 101

electricity, 12, 114, 130
    AC, 116
    batteries, 115
    conservation, 128
    DC, 116
    grounding, *124*
    homemade, 180
    production, *119,* 129
    savings, 129
    solar, 111, 127
    storage, 115, 129
    surplus, 111, 112
    usage, 102, 103, 107, 108, 110, 116, 127, 205
emergency
    back-up, 116
    medical bills, 169
    room (department), 168
emotions, 189
    emotional aftermath, 169
    attachment, 183
    disengagement, 183
    drive, 184
    interaction, 7
    rest, 185
    skill, 183
    turmoil, 185
emotionally strenuous, 185
endophytes, 90
end-of-life, 155
energy
    as a resource, 145
    efficiency, 12
    efficient, 82, 83, 102, 115, 116, 127
    hog (guzzling), 101, 127
    personal, 19, 27, 28, 95, *158,* 166, *175,* 179, *187,* 188
    self-sufficiency, 14, 20, 55, 101-130
    solar, 23, 55, 82, 115, 127
    star, 102, 103, 115
    storage, 114
    supply, 180
    sustainable, 13
entertainment, 160, *183*
enthusiasm, 20, *158, 163,* 164, *175,* 175, 178

environment, 15, *143,* 181, 190
environmental problems, 7, 183
environmentally responsible, 10, 126
environmentalists, 143
equipment, 8, 12, 19, *53,* 54, 71, 79, 106, 140, 143-145, 152, 153, 158, *161,* 172, 182, 205
erosion
    control, 209-214
    role of soil aggregates, 61, 148
establish
    food growing, 11
    hedgerow, 165
    homestead, 11, 44, 52, 53, 56
    pastures, 91, 210
    priorities, 9
    silvopasture, 49
established homesteaders, 53, 55, 56
establishment
    phase, 53, 54, 55, 56, 186
    projects, 22
Europe, 17, 144
European
    -American farmers, 17
    cheeses, 76
    scythe, *106*
evaluate
    choices, 9,
    income, 169
    priorities, 21-29, 180
    progress, 54, 55
    projects, 29, 160, 173, 176, 181
    reevaluate, 49, 176
    time, 160
evaluation process, 178, 181
evolution, 8
experimentation, 99
experiment(s), 172
    cheesemaking 76, 78
    cooling, 83,
    egg preservation, 75
    failed, 5
    feeding, 86
    filtration, 133-135, 137
    greywater, 131
    growing, 58, 65, 68, 71, 140
    solar, 55, 113

F

failure, 65, 188
fall (autumn), 76, 186, 187
    growing season, 67
    harvest, 69, 187
fans
    12-volt, 113, 205
    attic, 83, 107, *107, 109*, 111, 206
    box, 107, 113, 205
    ceiling, 107, 111
    cooling, 124, 126
    Ecofan, 111, 205
    solar, 83, 107, 111, 206
    vent, *109, 121*, 206
farm calendars, 17
    manuals, 17
Farmer Boy, 180
farmers, 135, 190
    almanacs, 17
    career, 18
    European American, 17
    natural, 99
    personal journals, 17
farmers cheese, 77, 204
farming, 14, *18*, 19
    conventional, 12
    venture, 18
    natural, 68, 90
farmsteads, 33
fatigue, 185
fence/fencing, 11, 18, 22, 35, 36, 45, 52-54, 70, 99, 156, 164, 166
    cattle panel, *94*, 165
    chicken yard, 55
    electric, 47, *49, 52*, 93, 111
    garden, 34
    lines, 35, 36, 44
    maintenance, 179
    post hole digger, 145
    repair, 22, 24, 155
    welded wire, 97
fertilizer, 90, 93, 147,
    organic, 206
    synthetic, 151
    whey as, 78
fescue 90
feta cheese, 76, 77, 204

Fiebig, Alan and Arlene, 111, 116
field
    crops, 12, 36, 146, 150, *150*, 151
    grain, 163
    hay, *21, 43*, 144, *151*
    peas, 212
filter
    cloth, 134
    inline, 134
    rainwater, 22, 55, 132-135, 172
    tank top, 135
filtration
    bed, 137
    problems, 51
    rainwater, *134*, 172
    soil, 61, 148
filtering system, 135
financial
    aid, 169
    consideration, 126
    reach, 111
    resources, 8, 182
    security, *11*
    situation, 169, 183
firewood, 4, 12, 16, 22, 23, 153, 157, 158
    hauling, *161*
    storage, 139
Five Acres and Independence, 135
fixer, nitrogen, 211-212
fixer-upper, 176
flail, 68, 69
flooding, 49, 51, 68
food
    budget, 126, 172
    canned, 80
    convenience, 82
    dehydrated, 25, 73, 80
    dry, 73, 80
    familiar, 83
    foraged, 66, 95
    forest, 165
    fresh, 72
    frozen, *74*, 77, 78, 80-82
    grocery store, 8
    homegrown, 12, 16
    homemade, 82
    industry, 80

lacto-fermented, *74*, 171
legalism, 84
loss, 113
mandala, *58*
plots, 12
preferences, 84
preservation, 18, 58, 72-79, 83, 126, 172, 179, 186, 187
production, 22, 54, 57, 58-72, 83, 140, 172, 179
scraps, 33, 97
seasonal, 17, 69
self-reliance, 20, 84
self-sufficiency, 12,
   animals. 85-99
   people, 57-84
storage, 12, 25, 58, 73, 79-83, 116
super-powered, 88
sustainability, 13
"food first," 24, 113
FoodSaver attachments, 73, 206
forage, 49, 187
   annual, 89-91, 209-214
   based feeding, *86*, 89, 95, 99
   deer mix, 89
   diversity, *89*, 93, 208, 209-214
   foods, 66
   grazing, 92
   growth, 47
   hedgerow, 54
   intensive grazing, 92-93, 151
   mob grazed, 92-93
   nutrients in, 88
   pasture rotation for, 207
   perennial, 90, 151, 209-214
   planting, 152
   polyculture, *89*, 209-214
   poor, 92
   problems growing, 54, 89, 90, 92, 163, 165
   products, 86
   rest, 92, 207, 208
   trampled, 92, 208
foragers, 33
forbs, 209, 213-214
Ford Powermaster 861, *145*
forest garden hedgerow (see also 'permaculture hedgerow'), 34-36, 45, 67-68, 94
fossil fuels, 106
free-range, 54, 95, 465
freezer, 36, 74, 80-82, 113, 121
   cheese, 77, 78
   chest, 82, 127
   chest versus upright, 116
   compressor, 129
   energy usage, 114, 116, 130
   ice cream, 195
freezer/refrigerator thermostat, *127-128*, 206
freezer to fridge conversion, *82*, 116, 127, 204
French
   bread, 84
   press coffee maker, *105*
fridge/refrigerator, 25, 75, 80-82, 113
   cheese cave, 76
   chest, 82, 127, 128
   chest versus upright, 116
   conversion, 116, 127, 204
   energy usage, 114, 115, 127
fruit trees, 12, 44, 57, 67, 138
   list of, *45*
frustration, 12, 18, 54, 76, 85, 86, 90, 160, 164, 171, 172, 177, 185
Fukuoka, Masanobu, *58*, 68, 90
Fukuoka style planting, 91, *91*, 95, 150, 151
functional
   benefit, 181
   part of ecosystem, 190
   soil, 51
fuses, 115
fusion, bone, 169

G

garden, 2, 12, 14, 18, 23, 44, 140, 150, 151, 158, 163, 164
   African keyhole, *142*, 204
   beds, 55, 59, *60*, 62-64, *65*, 131, 156
   books, 147
   canning, 150
   cart, *152*, 205
   climate challenges, 61, 65, 138, 142

(garden continued)
    controlling weeds, 12
    fall, 72
    fence, 34
    forest hedgerow, 34, 35, 45, 67, 93
    greywater, 138
    herb, 53, 150
    hügelkultur, 61, *62-64*, 65, 131
    kitchen, 150
    main, 57
    olla, 142, 205
    root crops in, 24
    scraps, 33, 98, 159
    seeds, 81
    soil building, *62-64*, 146
    suburban, *116*
    swales, 61, *62-64*, 65, 131
    tilling, 58
    water for, 61, *62-64*, 67, 131, 132, 136, 137, 142
    winter, 72, 80
    wiregrass in, 59, 90
    year-around, 69, 79
gasoline, 12, 106
gate(s), 35, 47
    arches, 46, *49*
    chicken, 54, *165*
    hayloft, *41*
    electric fence handles, *49*
    latches, 23
    locations, 45
    maintenance, 23
    milking room, *176*
generator, 126, 127
genetic potential, 88
genetically modified, 86
globe onions, 66
glomalin, 61, 148
glucose, 147
goals
    analyzing, 188
    balance, 187
    breeding, 177, 178
    dietary, 84
    energy, *101*, 116, 130
    feeding livestock, 85-86, 99
    food preservation, 72, *75*

    food storage, 80
    homestead, 27, 171, 177, 178
    keeping manageable, 29
    pasture improvement, 91-92, 207-208
    primary, *9, 21*, 22, 176, 178, *189*, 190
    reassessing, 9-20
    relevance, 10
    seasonal living, 16-18
    self-reliance, 1, 10, 19, 20, 23, 53, 55, 56, 177, 181
    self-sufficiency, 54, 57, 61, 83, *85*, 130, 184, 189
    self-supporting, 18-20
    simplicity, 11-13
    stewardship, 14-15
    sustainability, 13-14
    visualizing, *31*
    working toward, 12, 14, 55, 172, 173, 175, 177, 180, 183, *184*
goat(s)
    barn, 3, 22, 25-27, *28*, 28, 32, 34-36, *37-44*, 55, 90, 132, 155, *156*, 168, 171, 176
    breeding, 69-70
    corral, 36
    dairy, 47, 69
    digestive system (rumen), 88
    feed/feeding, 41 69, 71, 86, *86-88*, 88-90, 93, 94, 95, 181
    fences, 24, *45*, 47, 155
    grazing, 47, 92-93, 95, 207-208
    hay feeder, *41*, *43*
    head stalls, 41
    hedgerow, 54, 94
    kids, *173*, 178
    Kinder, 19, 69, 177, 204
    milk, 80, 177
    Nigerian Dwarfs, 69
    pasture, 46, 48, 49, 91, 207-208
    polio, 167
    problems, 24, 164, 167
    sales, 19
    self-supporting, 19, 95
    shed, 3, 7
    shelter, 33
    water, 137, 140

yard, 157, *159*
God, 8, 190
GoFundMe, 170
grain
    by-products, 86
    chickens, 95, 98
    ducks, 95
    crops, 26, *150*
    field, 163
    goats, 86, *86*, 88, 91, 95
    grinder (mill), 104, *104*, 205
    growing, 57, 68, 90, 99, 144, 145 152
    harvesting, 172
    no-till drill, 145
    pasture, 209-211
    processing. 68
    production, *150*, 151, 152
    sprouting, 88
    storage, 81, 128
    threshing, 68, 205
    vacuum canning, 73, *74*
Grandpappy's Official Website, 80
Grandpappy's Recipes for Hard Times, 71
grass(es)
    annual, 91, 209-211
    Bermuda, 58, 90
    chickens, 98
    clippings, 61
    fescue, 90
    goats, 93
    hedgerow, 45
    lawn, 58
    long-leafed, 209
    monoculture, 90
    native, 166, 209-211
    pasture, 58, 89, 91
    perennial, 91, 209-211
    polyculture, 209-211
    seed (in compost), 158
    small grain, 99, 209-211
    sorghum-sudangrass, 90, 210
    wiregrass, 58 59,*59*, 65, 90
grass-fed, 95
grassland, 93, 99, 209
grateful, 173, 186

grazing
    animals, 99, 180, 207
    beds, 97, *98*, 137, *141*
    intensive, 46, 92, 95, 151, 208
    mob, 92, 93, 208
    poor, 90
    rotation(al), 46, 90, 92, 151, 181, 207
    silvopasture, 49
    winter, 89
    year-round, 90
greenhouse, 23, 32, 34, 36, 44, 48, 54, 55, 67
green manure, 210-212
greens
    chickens, 98, 146
    fresh, 81, 98
    greywater, 138
    salad, 67
    goats, 71, 88, 99, 146
    winter, 80
greywater
    filtration bed, 137
    irrigation, 131, 138, *138*
    laundry, 137
    pH, 138
    precautions, 138
    problems with, 138
    recycling, *131*
    safe products, 138-140
    storage, 138
grid
    electrical, 101, 116, 130
    off-grid, 79, 111, 126, 129, 130
    tied, 111-113
grill, 108, *109*
grocery
    bill, 83
    store, 8, 12, 83, 126
ground, 48, 49, 56, 59, 61, 91, 93, 148, 149, *180*
    breaking, 34
    breeding, 138
    cover, 157, 166, 211
    higher, 51
    ivy, 91
    saturated, 49
    winter storage, 79

grounding
    earth, *122, 124*
    electricity, 124
    rod, *124*, 125
    wire, *124*
growing season, 17, 34, 67, 164
guinea fowl, 164, 166
guinea hogs, 33, 98, 203
gutters, 22, 134, 136

H
habit, *104,* 129, 160, 185
Halloumi, 76, 77, 204
hand
    powered, 104
    sanitizer, 140
Handy Farm Devices and How To Make Them, 135
hardwoods, 154, 155, 158
Harbor Freight Central Machinery sawmill, 26, 206
harvest, 66, 72, 86
    food first, 24
    grain, 172
    hay, 211
    meat, 33, 69, 187
    season, 16, 17, 28, 186, 187
    soil nutrient, 147
    summer, 59
    wheel, *58*
hawks, 98, 166
hay
    chute, *41, 43*
    feeder, *41, 93,* 205
    feeding, 51, *86,* 88, 89, 93, 95, 146
    fields, 144, *150,* 151
    growing, 12, 36, *89,* 99, 152, 209-214
    harvesting, *106,* 165, 172
    rake, *152,* 205
    seed source, 206
    waste, 90, *152*
    wheat, 90
hayloft, 15, *39, 42, 43*
    doors, 3, *44*
    roof, 136
head stalls, *41*

health, 15
    care, 170
    chickens, 54, 98
    ecological, 15
    goats, 95, 207
    insurance, 170
    plants/forage, 15, 56, 91, *146,*147, 207
    problems, 166, 167
    soil, 15, 51, 56, 65, 146, 208, 209
heater
    electric, 101, 102
    water, 101
    wood, 111
hedgerow
    forest garden, 34-36, *45,* 67-68, *94,* 163
    permaculture, 54, 150, 163-166
herb(s)
    garden, 11, 53, 68, 150
    hedgerow, 54, 166
    pasture, 89
    potted, *116*
    salves, 170
    vacuum canning, 73
herbal mineral mix, *88,* 99
herbed olive oil, 76, 204
herbicides, 12
hickory for grilling, 109
hobbyist, 26, 86
Holistic Management, 92
Holzer, Sepp, 61
homegrown
    diet, 83
    food, 12, 17
    hay, 43
    ingredients, 86
    recipes, 191-200
    vitamins & minerals, 88
    wheat straw, 90
homestead(ing)
    burnout, 185, *186*
    changes, 3-7
    day-to-day, 19
    ecosystem, 7, 56, 190
    establishment, 11, 53, 55, 56, 186
    foundation, 55
    goals, 9-20, 27, 171, 176, 177, 178,

  189, 190
 grown, 95
 ideas, 175
 journey, 1, 172
 library, 135
 lifestyle, 10, 16, 172
 livestock, 1, 33, 54, *85*, 88, 177
 making a living from, 18-20
 master plan, 31-52
 misconceptions, 9, 179, 180
 priorities, 21-29
 problems, 85, 160, 163, 166-173, 179, 185
 progress, 18, 32
 recipes, 191-200
 relationship to, 32
 resources, 143-162
 time-management, 160, 166, 181, 182
 transition, 53
 vs. job, 176
 work-smarter-not-harder, 97, *158*, 187
home-time, 177
honeybees, 34, 57, 70
hoop house (polytunnel), 67, *68*
hops
 vines, 71
 yeast, 71
hospital, 168-170
house
 back porch kitchen, *109*
 chicken (coop), *34, 97*
 doors, *102*
 duck, *141*
 electricity usage, 101, 106, 107
 energy efficiency, 102, 103, 107
 heating, 111
 roof runoff, 132
 repair & refit, 12, 18, 176
 then & now, 3, 4, 6
 upgrades, 18, 176
 water, 131
 windows, 22, 54, 102, 103
household
 budget, 117
 cleaners, 140

 members, 169-170
 spending, 172
How To Bake Without Baking Powder, 78
How To Preserve Eggs, 75
hügelkultur, 61
 swale beds, 61, 63-64, *65*, 131
human, 95, 137, *138*, 143
 consumption, 68
 health issues, 168
 invention, 162
 life-meaning, 10
 lifestyle, 19
 nature, 15, 56, 183, 187, 189-190
 purpose, 55-56, 190
 qualities, 20
 species, 7-8
humankind, 8
HVAC, 101, 111, 129
hydrogen peroxide, 138

I

ice cream, 81, 195
industrialized
 models, 88
 production, 86
industrial
 chipper, 157
 revolution, 189
infiltration
 problems, 51
 water, 148, 213
information, saving 10
infrastructure, 11, 12, 17, 53
 projects, 55
ingredients
 feed, 86
 homegrown, 86
 soap, 140
 substituting, 84
insect
 as chicken feed, 97
 beneficial, 211-214
 control, 81, 90
insulation
 house, 83, 102, *103*, 107
 soil, 67

INDEX 229

insurance
    car, 169, 170
    grid-tied solar, 112
    health, 170
intensive rotational grazing, 46, 92, 151, 208
interruptions, 164, 181
intuitive, 187
invasive, 59, 90, 212
inverter, 115, 116, *122*, 125, 127, *128*, 130
irrigation, greywater, 131, 138, *138*

J
Jerusalem artichokes, *45, 87*
job(s),
    retirement, 160
    to make a living, 11, 18, 164, 171, 172, 177, 180, 189
Judy, Greg, 93

K
Keeping Bees With a Smile, 70
kefir, 76, *78*, 84, 171
keyhole garden, 142, 204
Kill-a-watt meter, 114, 127, 205
kimchi, *74*
Kinders, 19, 69-70, 95, 177, 204
kitchen
    acids, *77*
    back porch, *109*, 187
    cart, *109*, 206
    garden, 150
    heating, *108*
    pantry, 79
    projects, 22
    scraps, 33, 98, 99, 158, 159
    storage, 80-82
    tools, *104*
    wind-up clock, *105*
KitchenAid mixer, 104, *104*
knee braces, 156
knowledge, 53, 54, 56, 86, 144, 150, 166, 167

L
lacto-fermentation, 73, *74*, 77, 81, 171
latches, 23, *103*

laundry, 19, 172, 185
    greywater, 137, 137
    outdoor, *139*
    plunger, 205
    products, 138
    rainwater for, 137
    tubs, *139*, 204
lawn, 24, 58
lawn mower, 68, *152, 161*
    wagon, *161*
Layens beehive, 70, 204
lay-in lugs, *124*, 125
Lazutin, Fedor, 70
lead-acid batteries, 125, 126
learning curve, 1, 57, 86, 156, 164, 166
legumes
    feed, 86, *86*, 209
    nitrogen-fixing nodules, 148, *149*
    pasture, 89, 91, 211-212
leisure, 1, *11*, 15, 26, 160, 186
lettuce, 16
    Jericho, 66
    Lollo Bianda, 66
life
    changes
    choices
    control of, 2, 180
    course, *178*
    discouraging things, 173
    homesteading, 16, 160
    maintenance chores, 185
    mistakes, 162
    modern, 162, 180
    plant, 151, 209
    purpose, 7, 8, 10, 56, 189-190
    questions, 189
    satisfaction with, 183
    simple, 8, *11*, 11
    shelf, 80
    skill, 183
    span
        batteries, 125
        trees, 155
    unhappy, 162
    way of, 1, 3, 145, 180
lifestyle
    agrarian, 25, 186

changes, 2, 54, 55, 84, 101, 104, 106
choices, 10, 20, 162
consumer-dependent, 172
environmentally responsible, 10
fixed income, 1
habits, 185
hands-on, 20
homesteading, 85, 164, 172, 182, 186
modern, 180
pandemic, 2
routine of, 18
seasonal, 16-18, 187
self-reliant, 12, 111, 130
simpler, 8
transition, 180
typical, 3,
lime, 90, 147
limestone, 140
lithium batteries, 125
livestock
    feed, 19, 86, 88, 99, 145, 172
    fencing, 11, 35
    grazing, 90, 92, *92*, 93, 146, 207-208
    health, 166
    homestead, 1
    management, 179
    needs of, 56
    numbers, 187
    pasture, 209-214
    predators, 15
    self-supporting, 19
    shelter, 11
    sustainability, 13
    vaccines, 81
    water, 132, 137, 140
longevity
    appliance, 129
    nutritional, 80, 81
long-term
    energy average, 129
    needs, 27
    physical
        ability, 27
        energy, 27
    projects, 27

lumber
    curing, *156*
    from fallen trees, 26, 155
    hauling, 145
    home-milled, 118
    milling, 35, 36, 47, 155, 156, 172
    mini-mill, 155
    waste, 157
Lyme disease, 166

M
machinery
    eliminate use of, 93
    maintenance, 23, 106, 179
    repairs, 24
maintenance
    battery, 121
    chores, 23, 164, 185
    fence, 23, 179
    machinery, 23, 106, 179
    pasture, 89
    potential, 27
    projects, 22, 23, 181
    seasonal, 22, 187
    vehicle, 177
"make it do," 8
mandala, nature's food, 58
manure
    animal, 33, 61, 68, 90, 91, 92, 157, 177, *180,* 181
    green, 210-212
marine batteries, 125
master plan, 28, 31-36, 44-48, 52, 55, 164, 175
material wealth, *11*
MC4 connectors, 125
meat
    animals, 46-47, 52, 70, 96, 98, 177
    bones, 196
    cooking, 108, 109
    flavor, 33, 70
    grass-fed, 95
    harvesting, 33, 69, 187
    leftover, *74*
    preservation, 79, 81
    production, 88, 95
    protein, 69

(meat continued)
   smoking, 79, *109*
   storage, 75, 81
   sustainable, 13
menu planning, 83
microorganisms, soil, 56, 147, 148
mildew
   indoors, 107
   powdery, 78
milk
   goat breed, 69
   proteins, 78
   seasonality, 17, 69, 178
   storage, 25, 75, *129*
   surplus, 80, 98
   sustainable supply, 13
   preserving, 76
   production, 88, 177
   year-around, 69, 70, 177
milking
   chore, 47
   equipment, 140
   room, *39-41, 112, 132, 133, 137, 176*
milling lumber, 35, 36, 47, 155, 156, 172
mindset
   business *18*
   linear, 22
mineral(s)
   amendments, 91
   bone broth, 149, 171, 198
   bulk, 147
   deficiencies, 22, 89
   goats, *88*, 99, 209-213
   profile, 147
   soil, 61, 95, 147, 149
mini-mill, 155
mob grazing, 92, 93, 208
models
   ecological, 183
   economic growth, 183
   grass-fed, 95
   industrialized production, 86, 88, 89, 90
   nature's, 90, 190
   pasture rotation, 207-208
   resource, 143
   rest, 186

modern
   calendar, 17
   conventional farming, 12
   culture, 8
   energy, 130
   feed, 88
   life(style), 162, 180
   point of view, 8
   production model, 89
   thinking, 7
   times, 186
   social system, 56
   world, 1
modified
   Fukuoka method, 90, *91,* 95, 150, 151
   genetically, 86
money, 2, *11,* 19, 20, 126, 147, 155, 157, 162, 175, 183, 188, 189
   reality of, 177
   time and, 177-182, 187
   time versus, 26-27
monoculture, 90
mortgage, 19, 169, 170, 176
Mosquito Dunks, 137, 206
mosquitoes, 136-137
moths, pantry, 73, 81
motivation, *163*
motive(s)
   for solar, 126
   suspicious of, 170
   underlying, 19
   well-meaning, 14
mower
   lawn, 68, *152, 161*
   sickle bar, *106,* 145
mowing
   lawn, 24
   pastures, 151, 208
   production areas, 152
mozzarella, 76, 77, 204
mud control, 157
mulch, 12, 51, 146, 151
   and chickens, 54, 95, 165
   and wiregrass, 59, 65
   barn bedding, 91
   cardboard, 64

living, 212
pastures, 151
plant residue, 149
production areas, 152
raised beds, 59
straw, 68, 90
to feed soil organisms, 148
winter root crops, 25, 79-80
wood chip, 157
Muscovy
 ducks, 48, 70, 96, 141, 166
 meat, 70, 96
mycorrhizal fungi, 56, 61, 147, *148*, 148, 150, 158

N

National Food Processors Association, 80
natural
 beekeeping, 70, 204
 behaviors, *97*
 cheesemaking, 76
 creation, *8*, 189, 190
 diet, 181
 ebb and flow, 186
 ecosystem, 99
 elements, 55
 farming, 68, 90, 99
 light, *42*
 order, 166
 pattern, 187
 processes, 14
 resources, 143
 rhythm, 18
 shampoo, 139
 system, 190
 tendency, 183
 ways, 12, 180
 woodland succession, 155
 world, 7, 8, 56
nature
 and humans, 14, 56, 187
 and technology, 8
 and urbanization, 8
 balance, 15
 closeness to, 12
 culling, 95

Food Mandala, *58*
 model, 90
 paradigm, 56
 partner with, 20
 plan, 56
 point of view, 14
 success, 15
 trusting, 95
 way, 15, 149
needs
 and wants, 182-184
 assessing, 21, 25
 changing, 55
 discerning, 56
 focusing on, 185
 fundamental, 20
 livestock, *34*, 56
 long-term, 27
 meeting, 10, 12, 14, 19, 56, *85*, 96, 144, 158, 176
 of others, 143
 plant, 56, 147
 simultaneous, 166
 soil, 56
new
 normal, 2
 year, 17
nickle-iron batteries, 125
Nigerian Dwarf goats, 69
nitrogen
 atmospheric, 148
 compost ratio, 149
 fixing bacteria, 148
 fixing, *149,* 209, 211-212
 recycler, 209
 scavengers, 209-211, 213-214
 sequesters, 212, 213
 soil, 147-149
 sources, 157
no-poo, 139
Norwegian brown (whey) cheese, 78, *78*, 204
no-till
 cover cropping, 149
 gardening, 58, 145
 grain drill, 145, 149, 152
 planting, 152, 211

nutrient(s)
    content, 80
    cycling, 61, 148
    feed, 86, *88*
    gap, 83
    plant, 147
    requirements, 95
    scavengers, 213
    soil, 51, 61, 147, 148
    transportation, 150

O

obstacles, *18*, 18, 160, 183
offal, 99
off-gassing, battery, *121*
off-grid
    meat preservation, 79
    electricity, 111
    system, 126, 204
old-time ways, 79
olive oil, 76, 77, 83, 204
olla, *142*
one step at a time, 172
One-Straw Revolution, 90
onions, 66, 81
opossums, *15*, 166
orchard, 53
    grass, 210
organic
    fertilizers, 206
    gardening books, 147
    matter, 51, 61, 64, 65, 148, 211
    olive oil, 77
    production, 137
    soil amendments, 147, 206
orthographic effect, 61
outbuildings, 18, 44, 54
    chicken coop, 7
    goat shed, 7
over-the-road, 18, 169, 176
owls, 166
owner-operator, 177

P

paddock(s), 34, 52, 54, 93
    grazing, 207-208
    resizing, 47, 52,
    rotation, 47, 92
    seeding and mulching, 95
    testing, 147
pandemic, 2
paneer, 77, 204
pantry
    closing off, 111
    improvements, 83
    moths, 73, 81
    refrigerator, 115, 127
    roof leak, 22, 24, 25
    size, 79
    storage, 25
    temperature, 25, 76, 80-82
parallel wiring, *119, 122*, 125
parasites, 167, 207
pasture
    annual forage, 89-91, 209-214
    chickens and, 54
    diversity, 89, 90, 93, 208, 209
    drainage, *50*, 51
    establishing, 12, 53, 144
    forage, 49, *86*, 89
    forbs, 212-214
    forest, 49
    grass, 58, 90, 209-211
    health, 56, 178
    improvement, 27, 95, 146, 147, 166
    legumes, 91, 211-212
    maintenance, 89, 90
    maps of, *45, 150*
    monoculture, 90
    perennial, 90, 91
    planting, 89, 91, 94, 150
    polyculture, *89*, 209-214
    remineralizing, 146-147
    rethinking, 35, 36, 150, 163
    rotation, 46, 52, 90, 207-208
    seed(ing), 55, 91, 93, 206
    silvo, 49
    soil building, 146, 151
    subdividing, 34, 46, 47, 93
    sustainable, 90
    versus packaged feed, 88, 92
    weeds, 91, 163
    winter, 90

patterns
    livestock,
    rest, 185-186
    rhythmic, 17
pay-as-you-go, 117, 172
pectin, 197-198
perennial
    forages, 90, 151, 209-214
    grasses, 91, 209-211
    pasture
pergola, 137
permaculture
    hedgerow (See also "forest garden hedgerow'), 54, 150, 163-166
    earthworks, 61
pH, soil, 90
philosophers, 189
philosophy, 86, 88
photo journal, 172
photosynthesis, 147
physical
    ability, 27
    body, 185
    energy, 27
    interaction with the world, 7
    rest, 184-185
    work, 185
pickling lime, 90
pigs, 33, *33*, 35, 36, 52, 77
    American Guinea Hogs, 33, 98, 203
    piglets, 33, 36, 98
    problems, 99, 155, 164, 166
pine
    bark beetle, 155
    milling, 155, *157*
    trees, 26, *45*, 49, 153, 154, 155, 158
pioneer species, 163, 155
plan
    B, 130
    master, 28, 31-36, 44-48, 52, 55, 164, 175
planning
    beekeeping, 70
    for drought, 61
    goals and, 55-56
    location, 25

    meal/menu, 75, 83
    project, 17
    sessions, 187
    shopping trips, 160
    stages, 176
plastic, 8, *81*
play, 160, 184, 185
plow(ed, ing), 12, 56, 90, 145, 153, *180*
pole saw, 145
political crossfire, 10
politicians, 190
politics, 185, 190
polyculture, *89, 90*, 209-214
pond,
    for ducks, 70, 137, *141*
    location, 32
    seasonal, 51
potatoes
    planting, *180-181*
    solar cooking, 108
    storage, 81
    sweet, 25, 66
potted plants, *116*, 140
poultry
    predators, 15, 166
    yard (see also chicken yard), 70, 95, 98
powdered
    eggs, 75
    limestone, 140
power
    battery, 115, 129
    DC, 113
    electrical, 12, *106*, 116
    off-grid, 111
    outage, 113, 126
    solar, 82, 107, *110*, 113, 115, 127
    strip, 104, 130
    struggle, 10
    super, 88
    tools, 106
    will, 55
    wind, *110*
Powermaster 861, *145*
power take-off (PTO), 145, 157, *158*
predation, 34, 166
predator attacks, 166

predators, 15, 166
Preparedness Advice blog, 114
Prepper's Livestock Handbook, 164, 158-159, 187
Prepper's Total Grid Failure Handbook, 111, 113
preservation
    department, 75
    equipment, 79
    food, 18, 58, 72, 126, 172, 179, 186, 187
    methods, *75,* 79
    process, 79
    supplies, 79
    techniques, 73
Preserving Food Without Canning or Freezing, 76
primary goal, 9, *21,* 22, 176, 178, *189,* 190
primost, 78
prioritize
    goals, 175
    projects, 22, 24, 27, 28, 29, 113, 137
    sessions, 164
priority, priorities, *9*
    establish, 9
    evaluate, 180
    goals, 99, *160*
    project, 35
    reevaluate, 21
    seasonal, 187
    setting, 18, 31
    top, 24
    water usage, 137
probiotic(s), 171, 195
problem(s)
    budget, 115
    critter, 34, 51, 54, 55, 70, 165, 166, 167
    drainage, 51
    environmental, 7-8,
    farming, *18*
    fencing, 99
    financing, 177
    food storage, 80-82
    gardening, 67
    greywater, 138
    health, 166, 168, 170
    homesteading, 21-22, 163, 172, 175, 185
    house, 106-107
    human aspect, 7-8, 15, 56
    lack of knowledge, 85-86, 88, 89
    lifestyle, 185, *186*
    logistical, 96, 155
    machinery, 144-145
    pasture, 90, 91, 93, 95, 208
    rainwater storage, 136, 137
    soil, 14, 56, 62
    solving, 8, 10, 23, 55, 173, 179, 183, 190
    time, 160, 184
    trees, 99, 155
    weather, 49, 59, 60
production
    commercial, 90
    crops, 150-152
    food, 22, 58, 67-69, 84, 93
    grid-tied, 112
    industrialized, 86, 88, 89
    organic, 137
productivity, 15
profit, 19, 113, 170
prototypes, 172
Providence, 172
pumpkins, 12, 87, 180
Pump-N-Seal, 73, 206
purpose, sense of, 2, *8,* 10, 20, 160, 162, 188-190
PVC pipe, 134, 135, 136

Q
Q_{10} temperature coefficient, 80
quality
    commercial, 8
    forage, 86, 88-89, 93, 95
    hay, 88-89, 95, 99
    of stored foods, 80
Queso Blanco, *77*
quilt, barn, 3, *44*

R
rabbits, 52
rabbit trails, *175, 187*

raccoons, 166
rain shadow, *60-61*
rainwater
    collection, 12, 23, 61, 131, *133,* 137
    filters, 22, *134,* 135, 172
    pipe, 135
    problems, 134-135
    swales, 61, 64
    tanks (totes), 48, 55, *132-134, 136,* 137
raised beds
    garden, 59, 60, 64, 67
    greywater, 137
    hügelkultur, 61
rat race, *11,* 11
rats, 34, 166
recipes
    cooking, 193-200
    homegrown, 193-200
    homemade oxy bleach, 139
    homemade scrubbing powder, 140
refrigerator/fridge, 25, 75, 80-82, 113
    cheese cave, 76
    chest, 82, 127, *128*
    chest versus upright, 116
    conversion, 116, 127, 204
    energy usage, 114, 115, 127
refrigerator/freezer thermostat, *127-*128, 206
remote temperature sensor, 125, 206
renting equipment, 27, 157, *182*
repair
    carport, 139
    category, 22, 24
    equipment, 24, 106
    fence, 22, 155
    house, 6, 12, 22, 25
    projects, 20, 55, 164, 176, 179
    season, 16, 187
replacement cost (batteries), 126
resource(s)
    definition of, 143-144
    external, 10
    financial/monetary, 8, 182, 183
    foundational, 145
    homestead, 153
    information as, *10*
    management, 56, 143-153, 158, 179
    most important, 146
    non-renewable, 159
    project, 27, 28
    renewable, 158
    stewardship, 183
responsible/responsibility
    lifestyle, 10
    to self, 173
    use of resources, 56, 144
rest
    defined, 185
    emotional, 185
    forage, 208
    mental, 185
    models of, 186
    physical, 185
    seasonal, 186
    spiritual, 185
    work and, 185
retirement, 1, 19, 20, 160, 171
reusable canning lids, *72,* 206
revised, revision,
    calendar, 17
    master plans, 28, 48
    project list, 29
rhythm
    bio, 160
seasonal, 16, 17, 187
rice
    cooking/reheating, 108, 110
    growing, 68, 90
    storage, 74
ricotta, 78, 204
ripping chain, 155
rocket scientist, 148
roof/roofing
    barn, *39-40,* 132, 136, 137,
    carport, *139*
    dining room, 176
    metal, 3, 39, 40
    pantry, 22, 24, 25
    runoff, 132, 133
root
    cellar, 23, 25, 79, 81, 83
    crops, 89, 99, 138, 146, 151, 213, 214

(root continued)
    plant, 54, 61, 147, 148
    sugar beet, *71*, 214
    vegetables, 79, 81
rotational grazing, 46, 90, 92, 151, 181, 207
routine
    chores, 164
    daily, 24, 55, 169, 170
    lifestyle, 18, 29, 181
    maintenance, 22, 23, 27
    seasonal, *16, 27*, 187
rumen, 88
RV batteries, 125

S
Sabbath, 186
Salad Bar Beef, 92
Salatin, Joel, 92-93
salt
    brine, 76
    grain of, 1
    in soap, 140
    sea, 83
salting, 79
sand
    filter, 135
    soil particles, 148
sandy
    loam, 62
    soil, 61
satisfaction, *8, 183*
Savory, Allan, 92, 93
sawmill, 26, *26,* 28, 35, 47, 145, 155-158, 172, 176, 206
science, 8, 15, 190
scrape blade, 145
scrubbing powder, 140
    Bon Ami, 140
    homemade recipe, 140
scythe, *106,* 145, *151,* 206
seasonal
    activities, 16, 17
    chores, 19, 23, 54
    cycles, 18, 90, 187
    eating/foods, 17, 69, 75, 83
    electric usage, 101

flooding, 51
forage, 95, 208
lifestyle, 16, 17, 84, 164, 189
living, 16-18, 20
pond, 51
projects, 22, 24, 27
rhythm, 16, 29, 187
routine, 16, 27, 187
season(s), 16-18, 22, 29, 65
    growing, 24, 28, 34, 67, 151, 164, 186
    harvesting, 28, 69, 81, 186
    planting, 68, 90, 186
    rest, 186
Seasons of America Past, 17
seed
    annual, 91
    garden, 81
    going to, 67
    grass, 158
    mix, 90, 91, 149, 151, 204
    pasture, 55, 91, 93, 151, 206, 209-214
    perennial, 91
seeding, *91,* 95, 151, 152
self-confidence, 181
self-perpetuating, *66*
self-reliance, 144
    chicken diet, 181
    energy, 111
    food, 84
    goals, 1, 9, 10, 19, 20, 55, 190
    homestead, 176, 189
    projects, 22
    recipes, 193-200
    steps toward, 12, 172
self-sufficiency, 147, 177
    defined, 9-10
    energy, 20, *101,* 101, 106, 130
    feed, 86, 99
    food, 12, 20, 57, 65, 83, 83-84
    goals, 19, 20, 54, 85, 99, 130, 184, 189
    projects, 16
    resource, 12, 14, 143
    water, 20, *131,* 131
    working toward, 1, 21, 159, 187
self-supporting, 18, 19, 20, 96, 177

self-sustaining, 95
Sepp Holzer's Permaculture, 61
septic tank, 138
sequester
    carbon, 92, 208
    nitrogen, 149, 212, 213
series wiring, *120, 122*
shampoo, 139, 140
shed, 3, *7*, 36
shed light, solar, 42, 111, *112, 172*, 206
shelter, buck, 49, *50*
    goat, *23*
    livestock, 11, 53
short-term, 27
sickle bar mower, *106*, 145
silt, 148
silvopasture, 49
simple
    diet, 2, 11
    farming, 90
    homemade, 23
    life, 8, 11, *16*
    matter, 180
    methods, 153
    ways, 27
simplicity, 10, 12, 15, 20
Simplicity Model W, *144*
simplify, 12, 18
skills
    cheesemaking, *76*
    emotional, 183
    equipment, 144
    life, 183
    tools, 144
skunk guard, *70*
skunks, 70, 166
skylight (barn), *40*
Sloane, Eric, 17
slow
    answers, 14, 164
    cooking, 79
    growing, 155, 166
    healing, 171
    progress, 18, 164
smoking meat, 79, *109*
snakes, 166

soap
    greywater and, *138*, 139, 140
    nuts, 138
social
    conditioning, 160
    pressure, 11
    security, 19, 169, 171, 172
    system, 56
    upheaval, 10
society, 160, *183*, 189
sodium
    based chemicals, 140
    bicarbonate, 140
    carbonate, 140
    chloride, 140
    in greywater, 138, 140
    silicate (water glass), 75, 206
soil
    acidic, 90
    aggregates, 61, 148
    amendments, 91, 147, 206
    bacteria, 56, 148
    building, 14, 27, 46, 61, 68, 92, 131, 146-150, 152, 208-214
    carbon, 147-150, 153
    Cecil sandy loam, 62
    chickens and, 54-55, 98, 165
    clay, 61-63, 148
    compaction, 148
    deficiencies, 22, 89, 95
    earthworms and, 56
    feeding, 12, 149
    fertility, 149
    filtration bed, *137*
    fungi, 56, 61
    health, 15, 51, 56, 65, *146*, 208, 209
    improvement, 89, 147, 150, 151
    infiltration, 51, 148
    (micro)organisms, 56, 92, 147-149, 208
    minerals, 95, 147, 149
    moisture retention, 61, 92, 147, 208
    nitrogen 149
    nutrient cycling, 61, 148
    organic matter, 148
    particles, 61, 148

(soil continued)
    poor, 14, 65, 91, 146, 165
    sandy, 61
    sandy loam, 62
    tests, testing, 140, *146,* 147
    texture, 61
    Web Survey, 62, 204
    wiregrass and, 59
solar
    attic fan, 83, *107, 109,* 111, 206
    box fan, 113
    businesses, 114
    charging station, *101, 119,* 206
    clothesline, 110
    cost, 125-126
    electricity, 111
    energy, 55, 127
    experiment, 113
    fence energizer, 111
    oven, 79, 108, *108,* 111, 206
    panel array, *118, 122*
    panel rack, *117, 118,* 126
    panels, 4, 55, 113-115, *117,* 124, 125, 126
    powered, 107
    shed light, 42, 111, *112, 172,* 206
    system, 82, 127, 129
        paying for itself, 126-127
    usage fees, 113
"something is better than nothing," 2, 116
songbirds, 98
sorghum
    grass, 210
    syrup, 70, 71
sorghum-sudangrass, 90, 210
soul-searching, 6
sourdough, 71, 77, 78
specific gravity, 129
spring, *16,* 17
    cheesemaking, 76
    kidding, 69, 70, 187
    pasture, 151
    planting, *180,* 186
    season, 16, 186
squirrels, 98
starting (cranking) batteries, 115, 125

state of charge (SOC), 129
stewardship
    defined, 14
    goal, 10, 14-15, 20, 56, 162
    resource, 143, 183
    trees, 158
    sustainability and, 189
    water, 131
stocking density, 93, 208
storage
    cheese, 76
    electricity (battery), 115
    food, 12, 25, 58, 73, 79, 116
        challenges, 80, 84
        conditions, 80
        long-term, 78, 81
    space, *34,* 79
    techniques, 75
    temperatures, 82
stove
    coal, 176
    cook, 108, 111
    electric, 101, *110,* 129
    wood, 12, 22, 155
stovetop
    fans, 111
    percolator, *105*
strategies, soil building, 150
straw, 61, 68, 90, 91, 158
strawberry, 29
strut channel, *118, 119*
subjective, 27, 175, 181
success, 15, 19, 47, 65, 130, *158,* 160, 179, 180, 187
succession, 155
sugar beets, 70, 71, *71,* 214
summer, 16, 83
    air conditioning and, 106, 129
    cheesemaking, 76
    cooking, 108-109
    electricity usage, 107, 129
    food storage, 81
    garden, 67, 68, 186
    grass, 58, 90
    harvest, 59, 72
    heat, 16, 59, 61, 66, 80, 82, *106,* 121, 165

hoop house in, 67, 68
house temps, 10, 102, 107, 109
pasture, 90
projects, 187
power outages, 113
season, 16, 186, 187
weather, 59, 60, 89, *121,* 132, 137, 146
sunlight hours, 114
supply and demand, 144
sustainability
    defined, 13-14
    goals, 10, 15, 20, 189
    soil and, 14
sustainable
    agriculture, 13, 14
    economics, 13
    energy, 13
    food supply, 177
    hedgerow, 184
    pastures, 90
    word usage, 13
swales, 51, 61
    hügelkultur, 63, 65, 131, 151
sweet potatoes, 25, 66
swimming pool, *182*
symbiotic, 97, 147, 148
syrup,
    maple, 71, *78*
    sorghum, 70-71, 71
    sugar beet, 71, *71,* 204
system
    biological, 80
    chemical, 80
    consumer, 12
    digestive (goats), 88
    direct current, 115
    eco-, 7, 56, 99, 190
    economic, 9-10, 20
    feeding, 99
    gating, 47, 181
    grid-tied, 111-112
    immune, 207
    nature's, 15, 190
    no-till, 211
    off-grid, 126
    rainwater, 61, 131, 133, 135

root (plant), 147
social, 56
solar energy, 23, 82, 113, 114, 117, 129
    diagram, 119
    grounding, *124*
    paying for itself, 126-127
sustainable, 14
weather, 60
world, 190

T
table
    salt, 149
    saw, 168
tangible assets, 19
tank(s)
    overflow, 135
    rainwater, 48, 55, 138, 206
        barn, *132-136*
        carport, *139*
        house, *132,* 132, 133, 137
        problems, 136-137
    septic, 138
    stock, 141
tap water, 131, 138, 140
tarp, 25
tax
    incentives, 111
    rebates, 112
technological advancement, 8
technologists, 190
technology, 8, 15, 130, 182, 190
    sensible, 182
technophile, 182
technophobe, 182
thankful, 186
thermoelectric fan, 111. 205
thermostat
    fridge, 127, *128,* 206
    house, 106, *107*
think outside the box, *183*
tiller(s), 12, 144
tilling, 12, 56, 58, 59, 65, 90, 149, 153
timber
    construction, 37, 176
    curing, 37, 156
    milling, 145, 156

time
- as a gift, 160
- as a resource, 158-162
- as a tool, 160
- balancing, 27, 28, 84, 166, 179, 180
- daylight savings, 160
- frame, 184
- frustration, 160
- intensive, 75, 147
- leisure, 15, 160
- management, 160
- pressure, 160, 184
- rationing, 158, 175, 187
- saving, 26
- spending wisely, 160
- stewardship, 162
- trading for money, 162, 180-181, 182
- wasting, 160

time-consuming, 57, 176
toaster oven, 110
to-do list, 22, 24, 34, 162, *164*, 184
totes (see tanks)
tractor, *38*
- 2-wheel (walking), 144
- attachments, 145, 152
- chicken, 181
- cost, 145
- farm, 144, *145*, 172
- path, 47-49
- power take-off (PTO), 145, 157, 158

Tractor Supply Co., 132
trade-offs, 84, 102, 125, 129
transition, 1, 56, 83, 84, 180
trees
- as a resource, 145, 153-158
- chestnut, *45*, 54, 164
- falling, 24, 99, 155
- fruit, 12, 44, *45*, 57, 67, 138
- hardwoods, 154, 155, 158
- hedgerow, 54, 164-164
- pecan, 98, 164
- pine, 26, 153, *154*, 155, 157
- pioneer, 155
- renewable resource, 158
- silvopasture, 49
- trimmings, 93

waste, 28, 155, 157
woods, *45*
truck driving, 18, 19, 169, 171, 172, 177
turmoil, 185
two-wheel (walk-behind or walking) tractor, 144

U

uncertainty, 18, 168, 170
undersowing, 152
upright
- freezer, 116
- refrigerator, 127, 129

urbanization, 8
urine, 90, 91, 157
U.S., 71, 80, 135, 144, 170
U.S. Army study, 80
U.S. Energy Information Administration, 126
- usage
- electricity, 112, 116, 129
- fees, 113
- land, 150, *150*
- monitor, 114, 127, 205
- rainwater, 137

"use it up," 8
utility
- bills, 19, 169
- carts, 206
- company, 111-113
- room, 111

V

vacation, *186*
vacuum canning, 73, *73-74*, 204, 206
vegetable(s)
- best suited, 65
- garden, 12, 53, 150
- lacto-fermented, *74*, 81
- storing, 81
- roasted, 76
- root, 79

vehicle
- battery, 115
- repairs, 24

vent fan, 83, 107, *107*, *109*, 111, 206
vetch, 89, 212

veterinarian, 167
vinegar
    apple cider, 139
    bone broth, 149
    cheesemaking, 78
    cleaner, 140
vitamin(s), 88, 209
voltage, *120, 122*, 129, 130
vortex blender, *105,* 206

W

wagon, *161*
Waldo and Polly, 33, 36, 99
    walk-behind (walking or 2-wheel)tractor, 144, 144-145
warm climate pasture, 209
Warré
    beehive, 70, 206
    natural beekeeping, 70, 204
washing machine
    electric, 139
    greywater, 137
    off-grid, 139
wash tubs, *139,* 206
waste
    butchering, 33
    hay, 90
    lumber, 157
    plant, 68
    slaughtering, 33
    trees, 28, 155, 157
water
    as a resource, 145
    blackwater, 138
    conservation, *131,* 131, 140, 142
    drainage, 51, 62 141
    erosion, 61, 148
    filtration, 61
    filtration bed, 137
    glassing eggs, 75
    greywater, see greywater
    heater, 101
    hoop house and, 67
    infiltration, 148
    laundry, 138
    livestock, 140, 141
    municipal, 131
    rainwater, see rainwater
    rationing, 61, 132
    retention, 61-64, 131
    self-reliance goals, 20, 55
    self-sufficiency, 61, 131
    stewardship, 131, 140
    sustainability, 14
    tanks/totes, see tanks
    tap, 131, 138, 140
wealth, *11,* 20, 171, 180, 190
"wear it out," 8
weariness, 185
weather vane, 44
welding machine, 145
whey, 76
    cheeses, 78
    protein in, 78
    uses for, *33,* 33, 77-78
why am I here?, 160, 189
Wilder, Laura Ingalls, 180
willpower, 55
window(s)
    barn, *42*
    bay, 3, *6,* 176
    screening, 88, 136
    shading, *137*
    energy efficient, 83
    Energy Star, 103
    replacing, 27, 102, 137, 176
    single glazed, 102
wind-up clock, 105
winnowing, 68
winter, 16, 186, 187
    cold, 116
    cooking, 108
    garden, 72, 80
    forage, 89, 90
    grazing, 89
    hoop house in, 67
    layers, 75
    livestock management, 187
    milk supply, 70
    pasture, 90
    power outages, 113
    root crops, 25
    soups, 74, 82
    vegetables, 25, *66,* 79, 80, 87

(winter continued)
    weather, 25, 47, 80, 107,
    wheat, 68
wiregrass, 58-59, 65
wiring, solar
    diagrams
        battery bank, *122*
        whole system, *119*
    grounding, *124*
    parallel, *120*
    series, *120*
Wondermill, 104, 205
wood
    chipper, 157, *158*, 206
    chips, 61, 64, 157, *159*, 161, 171
    cookstove, 108, *108*, 111
    gasifier, 55
    grain, 155
    lot, 99, 154
    stove, 12, 22, 111, 155
woodland succession, 155
woods, 35, 47, 52, 99, 153-155, 158
    silvopasture, 49
woody annuals, 91
work
    and rest, 180, 184
    flow, 176
    horse, 145
    in progress, 32
    life's, 7
    load, 158, 181, *187*
    purpose in, 189-190
    routine, 29
    to rest ratio, 186
    versus play, 160, 184
    week, 19
"work smarter not harder," 12, *97*, *158*, *161*, 181, 182, 187
workshop, 3, 7, 36, *48*, 111
world
    modern, 1
    natural, 7, 8, 56, 189, 190
    new, 190
    saving, 190
    shaped by human hands, 189
    way, 8
wormer, 167, 207

worship, 186
wound care, 170-171

X
x-ray, 168-169

Y
Y_2K, 73
yard chipper, 69
yeast, 71
YouTube, 147, 173

Z
Zimbabwe, 93

Interested in More?

5 Acres & A Dream The Book
The Challenges of Establishing a Self-Sufficient Homestead

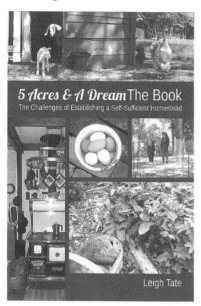

5 Acres & A Dream The Book is the first in Leigh Tate's 5 Acres & A Dream homesteading series. It begins with the years leading up to the Tates' 2009 purchase of a 1920s farmhouse on five acres in the foothills of Southern Appalachia. Spanning five years of their lives, this first book in the series shares how they defined their dream, developed property hunting criteria, and after several setbacks, found the right place. It describes how they set and prioritized their goals, and how they mapped out a plan. It shares their difficulties and obstacles, and what they've learned about energy, water, and food self-sufficiency for both themselves and their livestock.

Critter Tales: What my homestead critters have taught me about themselves, their world, and how to be a part of it

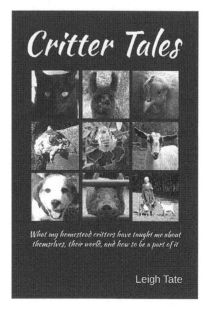

Although not a part of Leigh Tate's 5 Acres & A Dream series, Critter Tales nonetheless fills in the gaps of the Tate's homesteading story. Focuses on the many challenges of working toward sustainable livestock keeping. Contains amusing stories of their homestead critters, problems faced, successes and failures, and lessons learned. Offers an interesting insight into the different —and often conflicting—philosophies of managing livestock.

Prepper's Livestock Handbook
Lifesaving Strategies and Sustainable Methods for Keeping Chickens, Rabbits, Goats, Cows and Other Farm Animals

This prepper's resource by Leigh Tate focuses on sustainable and self-reliant methods for managing your land and your livestock. Covers livestock choices for homesteading, options for shelter and fencing, homegrown feeds and hay, breeding, birthing, veterinary care, and dairying. Learn pitfalls to avoid and how to keep things manageable. Learn simple, low-tech, off-grid ways to preserve eggs, milk, cheese, and meat. *Prepper's Livestock Handbook* will help you make the best livestock choices for your personal homestead needs and goals.

How To Bake Without Baking Powder
Modern and historical alternatives for light and tasty baked goods

Baking powder is a common kitchen staple, but did you know that you can create your own leavening power with common food items you already have on hand? *How To Bake Without Baking Powder* will teach you basic kitchen chemistry, so that you will never have to throw out expired baking powder again. Discusses the science of baking powder, plus substitutes for both cream of tartar and baking soda. 20 baking powder alternatives, 12 easy reference charts, and 54 modern and historical recipes.

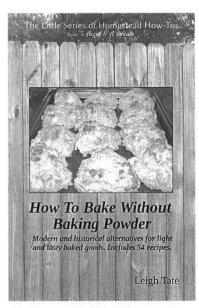

The Little Series of Homestead How-Tos

Leigh Tate's how-to eBook series is geared toward helping homesteaders fine-tune the knowledge and skills they need to work toward greater self-reliance and sustainability. Popular titles include:

How To Preserve Eggs: freezing, pickling, dehydrating, larding, water glassing, & more

How To Make an Herbal Salve: an introduction to salves, creams, ointments, & more

How To Mix Feed Rations With The Pearson Square: grains, protein, calcium, phosphorous, balance, & more

How-To Home Soil Tests: 10+ DIY tests for texture, pH, drainage, earthworms & more

How To Garden For Goats: gardening, foraging, small-scale grain and hay, & more

How To Grow Ginger: how to grow, harvest, use, and perpetuate this tropical spice in a non-tropical climate

How To Grow a Garden from Groceries: Fun ideas for the brave, the bold, the bored, & the downright curious

How To Make Amish Whitewash: make your own whitewash, paint, and wood stain

How To Compost With Chickens: Work smarter not harder for faster compost & happier chickens

And more. Visit http://kikobian.com/little_series.html for a complete list of current titles.

Made in the USA
Columbia, SC
19 January 2022